SPECIATION AND EVOLUTION OF THE PYGMY MICE, GENUS BAIOMYS

ROBERT L. PACKARD

ZINC READ

Published by
This edition copyright © 2024 by Zinc Read
Design and setting By Zinc Read
Email- zincread@gmail.com
ISBN-978-93-6277-671-6

CONTENTS

INTRODUCTION

Pygmy mice (*Genus Baiomys*) are the smallest cricetine rodents in North America. They occur from Nicaragua in Central America into the southwestern United States. The principal part of the geographic range of the pygmy mice lies in the Republic of Mexico. They are notably common in central Mexico, but are only locally common to the north and to the south, and then only in certain seasons.

Pygmy mice were first brought to the attention of biologists in 1887 when Oldfield Thomas described a diminutive species of cricetine rodent, *Hesperomys* (*Vesperimus*) *taylori*. The description was based on a specimen obtained by William Taylor from San Diego, Duval County, Texas. C. Hart Merriam (1892:70) described *Sitomys musculus* on the basis of specimens from Colima [City of], Colima, Mexico. Merriam (*loc. cit.*) mentioned that the two kinds of mice, *Hesperomys taylori* and *Sitomys musculus*, "in general appearance look almost precisely like the common house mouse (*Mus musculus*) but are still smaller and have shorter tails." He placed the two species in the genus *Sitomys*. Frederick W. True in 1894 regarded them as composing a distinct subgenus of *Sitomys, Baiomys*. According to True (1894:758), *S. taylori* and *S. musculus* possessed a different combination of characters (ascending ramus of mandible short and erect, condyle terminal, coronoid process well-developed, uncinate, and near the condyle, size small, tail short, plantar tubercles six, soles hairy) than either *Vesperimus*, or *Onychomys* (which had been considered as a subgenus of *Hesperomys* until 1889). In 1907, E. A. Mearns accorded *Baiomys* generic rank. Osgood (1909:252) treated *Baiomys* us a subgenus of *Peromyscus*, whereas, Miller, in 1912, regarded *Baiomys* as a distinct genus. Most recent students of North American mammals have followed Miller, but usually with reservations. Ellerman (1941:402) emphasized that the taxonomic position of the genus was uncertain, and wrote that *Baiomys* "... seems to be considerably distinct from *Peromyscus*, and may perhaps be a northern representative of *Hesperomys* or one of the small South American genera."

Only two comprehensive analyses of geographic variation and interspecific taxonomic relationships have been made; the first was by

Osgood (1909) who had fewer than a fourth of the specimens of *Baiomys* available to me; the second was by Hooper (1952a:90-97) who contributed importantly to understanding the relationships of the two living species in central Mexico. No attempts heretofore have been made to correlate and understand the relationships of the five fossil species to one another and to the living species assigned to the genus.

Six objectives of the following report are to: (1) list characters taxonomically useful in recognizing species and subspecies; (2) record amount of variation within and between populations; (3) correlate observed variations with known biological principles; (4) show geographic ranges of the two living species; (5) indicate relationships between fossil and living species of the genus; and (6) clarify the systematic position of the genus.

MATERIALS, METHODS AND ACKNOWLEDGMENTS

This report is based on the study of approximately 3,520 museum study skins, skulls, complete skeletons, and entire animals preserved in liquid. Most specimens examined were accompanied by an attached label bearing data on locality and date of capture, name of collector, external measurements, and sex. In addition, 49 fossil specimens referable to *Baiomys* were studied. Nearly two-thirds of the specimens were assembled at the University of Kansas Museum of Natural History; the remainder were examined in other institutions.

Specimens studied were grouped by geographic origin, sex, age, and season of capture. Individual variation was then measured in several of the larger samples of each living species and in measurable fossil material. External measurements used were those recorded by the collectors on the labels attached to the skins. Twenty cranial measurements employed in the past in the study of *Baiomys* and closely related cricetine rodents were statistically analyzed. The coefficient of variation was calculated for each of the 20 measurements in order to determine which varied least. In general, measurements having the least coefficient of variation were used in comparing samples from different geographic areas. Figure 1 shows the points between which measurements were taken.

Occipitonasal length.--From anteriormost projection of nasal bones to posteriormost projection of supraoccipital bone. *A* to *A′* *Zygomatic breadth.*--Greatest distance across zygomatic arches of cranium at right angles to long axis of skull. *B* to *B′* *Postpalatal length.*--From posterior margin of hard palate to anterior margin of foramen magnum. *C* to *C′* *Least interorbital breadth.*--Least distance across top of skull between orbits. *D* to *D′* [Pg 585] *Length of incisive foramina.*--From anteriormost point to posteriormost point of incisive foramina. *E* to *E′* *Length of rostrum.*--The distance in a straight line from the notch that lies lateral to the lacrimal to the tip of the nasal on the same side. *F* to *F′* *Breadth of braincase.*--Greatest distance across braincase, taken at right angles to long axis of skull. *G* to *G′* *Depth of cranium.*--The distance from the dorsalmost part of the braincase to a flat plane touching tips of incisors and ventral border of each auditory bulla. A glass slide one millimeter thick was placed on the ventral side of the skull. One jaw of the caliper was on the lower surface of the slide and the other jaw on the dorsalmost part of the braincase. The depth of the slide was subtracted from the total reading. *H* to *H′* *Alveolar length of maxillary tooth-row.*--From anterior border of alveolus of M1 to posterior alveolus of M3. *I* to *I′*

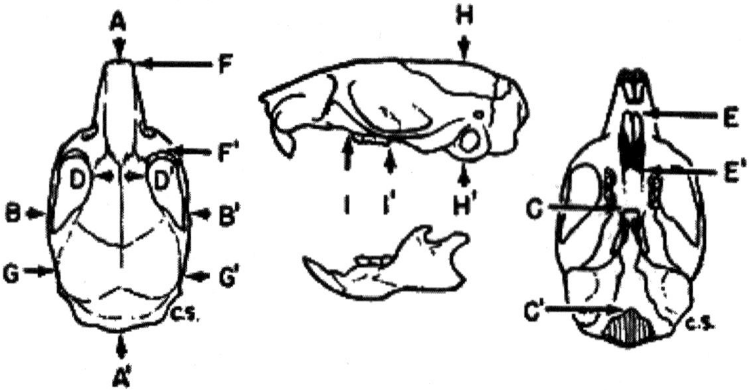

FIG. 1. Three views of the skull to show points between which measurements were taken.
Based on *B. m. pullus*, adult, female, No. 71611 KU, 8 mi. S Condega, Esteli, Nicaragua. x 1 1/3.

Capitalized color-terms refer to Ridgway (1912). Color terms without initial letters capitalized do not refer to any one standard.

The names of the cusps and ridges of the teeth (see Figure 2) are those suggested by Wood and Wilson (1936:389-390). Terminology of the enamel grooves and folds is that of Hershkovitz (1944:17) and Hooper (1952b:20-21).

Because secondary sexual variation was not significant (see page 597), both males and females of like age and pelage were used in comparisons of samples designed to reveal geographic variation.

The species are arranged from less to more progressive; the subspecies are arranged alphabetically.

In the synonymy of each subspecies, the plan has been to cite: (1) the name first proposed; (2) the first usage of the name combination employed by me; (3) all other name combinations in chronological order that have been applied to the subspecies concerned.

The localities of specimens examined are listed by country from north to south. Within a country, the listing is by state, beginning with the northwesternmost state and proceeding by tiers (west to east) to the southeasternmost state. Within a state of the United States, the listing is by counties in the same geographic order as described for states. Within any county in the United States, within any state in Mexico, and within any country in Central America, the listing of localities is from north to south. When more than one locality is on the same line of latitude, the westernmost locality is listed first. Marginal localities for each subspecies are listed in a paragraph at the end of each account. Each marginal locality is mapped by means of a circle. The circles are listed in clockwise order, beginning with the northernmost. When more than one of these localities lies on the same line of latitude, the westernmost is cited first. Localities not represented on the distribution maps, so as to avoid undue crowding of symbols, are italicized in the lists of specimens examined.

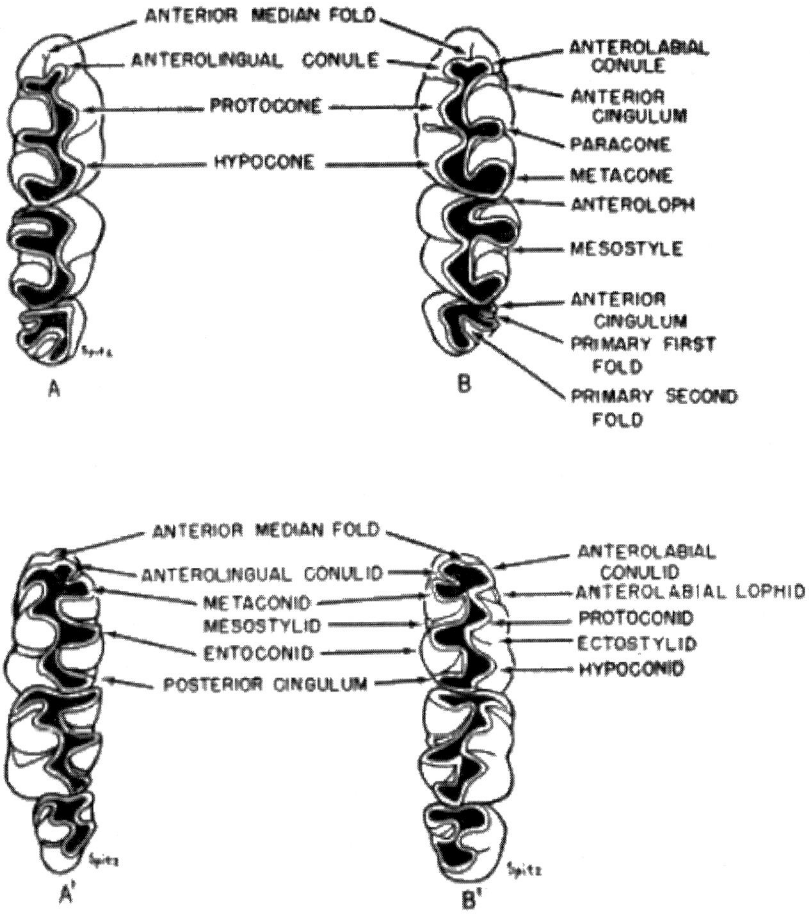

FIG. 2. Occlusal views of molars. x 13.

A.

B. taylori analogous, subadult, female, No. 28102 KU, 4 km. ENE Tlalmanalco, 2290 meters, Estado de Mexico. Right, upper molars.

B.

B. musculus musculus, subadult, male, No. 45456 USNM, Colima, Colima, Mexico. Left, upper molars.

A'.

B. taylori analogous, subadult, female, No. 28102 KU 4 km. ENE Tlalmanalco, 2290 meters, Estado de Mexico. Left, lower molars.

B'.

B. musculus musculus, subadult, male, No. 45456 USNM, Colima, Colima, Mexico. Right, lower molars.

The largest single collection of pygmy mice is in the University of Kansas Museum of Natural History, and, unless otherwise indicated, specimens cited in the taxonomic accounts beyond are there.

I am indebted to the following named institutions and persons for making specimens available for study:

American Museum of Natural History, G. G. Goodwin and R. G. VanGelder.

Carnegie Museum, J. K. Doutt.

California Academy of Sciences, Robert T. Orr.

Chicago Natural History Museum, Phillip H. Hershkovitz.

Cleveland Museum of Natural History (Collection now a part of Museum of Zoology, University of Michigan, W. H. Burt, E. T. Hooper).

Louisiana State University, Museum of Natural History, George H. Lowery, Jr.

Los Angeles County Museum, Charles A. McLaughlin.

United States National Museum (Biological Survey Collections), David A. Johnson, and Viola S. Schantz.

United States National Museum, Division of Vertebrate Paleontology, C. Lewis Gazin.

University of Arizona, E. L. Cockrum, and G. VR. Bradshaw.

University of California, Museum of Vertebrate Zoology, Seth B. Benson, and W. Z. Lidicker.

University of Illinois, Museum of Natural History, Donald F. Hoffmeister.

University of Michigan, Museum of Zoology, W. H. Burt, E. T. Hooper, and Claude W. Hibbard.

University of New Mexico, James S. Findley.

University of Texas, Frank W. Blair.

Texas A & M, Cooperative Wildlife Research Collection, W. B. Davis.

The Museum, Michigan State University, Rollin H. Baker.

University of Florida Collections, James N. Layne.

I am especially grateful to Professor E. Raymond Hall who guided me in my study and gave critical assistance with the manuscript. Additional appreciated suggestions were made by Professors A. Byron Leonard, Robert W. Wilson, Henry S. Fitch, Ronald L. McGregor, and fellow graduate students. For the illustrations, I am indebted to Mrs. Lorna Cordonnier, Miss Lucy Remple and Mrs. Connie Spitz. Mr. B. J. Wilks of the University of Texas, Department of Zoology, provided a number of living pygmy mice for study in captivity. Mr. J. Raymond Alcorn and his son, Albert, collected a large share of specimens of pygmy mice now in the University of Kansas, Museum of Natural History. My wife, Patricia, aided me in secretarial work and typing of the manuscript.

For financial assistance, I am indebted to the National Science Foundation when I was a Research Assistant, to the Sigma Xi-RESA Research Fund for a Grant-in-Aid, and to the Kansas University Endowment Association through its A. Henley Aid Fund, and the Watkins Fund for out-of-state field work by the Museum of Natural History.

PALEONTOLOGY OF THE GENUS

Five fossil species, all extinct, have been assigned to the genus and range in time from early late Pliocene (Saw Rock Canyon fauna of Hibbard, 1953:408) to Mid-Pleistocene (see Hibbard, 1958:25, who assigns the Curtis Ranch fauna to late Kansan or early Yarmouth).

I examined all known fossil material and compared it with Recent material. When the antiquity of the genus is considered, the degree of difference between the oldest fossil species and the two living species is much less than might be expected.

Baiomys sawrockensis Hibbard

Baiomys sawrockensis Hibbard, Papers Mich. Acad. Sci., Arts and Letters, 38:402, April 27, 1953.

Type.--No. 27506, Univ. Michigan; left mandibular ramus bearing m1-m3 and incisor; Saw Rock Canyon, early late Pliocene, XI member of the Rexroad formation, sec. 36, T. 34 S, R. 31 W, Seward County, Kansas (University of Kansas, Locality 6).

Referred material.--Univ. Michigan, Nos. 25781, 27503-27505, 28159-28165, 29708-29715, 31015.

Diagnosis.--Ramus of medium size to small for the genus; lower incisor broad, moderately recurved; diastemal region broad; anterior median fold between anterior labial conulid and anterior lingual conulid of m1 deep; primary first fold between anteroconulid and protoconid of m2 deep; cingular ridge (ectolophid) at entrance to posteroexternal reentrant valley (major fold, see Figure 2) between protoconid and hypoconid of m1 and m2; average and extreme measurements of lower molar row of eight specimens are, 2.65 (2.5-2.7).

Comparisons.--For comparisons with *B. brachygnathus*, see account of that species. From *B. rexroadi*, *B. sawrockensis* differs in: anterior median fold of m1 deeper; incisor narrower; diastemal region broader; coronoid process broader and better developed; cingular ridges (ectolophids and mesolophids) more pronounced in their development; incisors less proodont, more retrodont.

From *B. kolbi*, *B. sawrockensis* differs in: crowns of molars narrower; incisors less proodont; cingular ridges (ectolophids and mesolophids) of m1 and m2 more pronounced in their development.

From *B. minimus*, *B. sawrockensis* differs in: incisor less procumbent; masseteric ridge extending farther anteriorly; anterior cingulum of m2 slightly larger.

From *B. musculus*, *B. sawrockensis* differs in: over-all size of jaw and molar row less; diastema more acutely curved; incisors shorter; anterior median fold of m1 slightly deeper.

From *B. taylori*, *B. sawrockensis* differs in: m1 and m2 smaller; cingular ridges in m1 and m2 more pronounced; anterolingual conulid farther forward; incisors shorter, more proodont; molar teeth depressed, less hypsodont; diastemal region broader, more acutely curved; masseteric ridge not extending so far anteriorly.

Remarks.--*B. sawrockensis* is the oldest known pygmy mouse. The extreme development of the anterior median fold between the anterolingual conulid and the anterolabial conulid is regarded as a primitive feature in the pygmy mice. In this character, the Recent species can be traced back in time through *B. minimus* to *B. sawrockensis*. *B. sawrockensis* resembles *Calomys laucha* of South America in general conformation of jaw and tooth structure. The molars of *sawrockensis* are smaller than those of *C. laucha*, and the anterolingual conulid of *sawrockensis* is farther forward.

Baiomys rexroadi Hibbard

Baiomys rexroadi Hibbard, Amer. Midland Nat., 26:351, September, 1941; Hibbard, Contrib. Mus. Paleo., Univ. Michigan, 8(2):145, June 29, 1950 (part); Hibbard, Papers Mich. Acad. Sci., Arts and Letters, 38:403, April 27, 1953.

Type.--No. 4670, Univ. Kansas; left mandibular ramus bearing m1-m3, and incisor; Rexroad fauna, Locality no. 2, Upper Pliocene, Meade County, Kansas.

Referred material.--Univ. of Michigan Nos. 24840, 24851, 27493, 27496, 27501, 28862-28867.

Diagnosis.--Ramus medium in size for the genus; incisors small, proodont; anterior median fold of m1 slight; cingulum of all molars poorly developed; average and external measurements of lower molar row of seven specimens are, 2.7 (2.6-3.0).

Comparisons.--For comparisons with *B. sawrockensis* and *B. minimus*, see accounts of those species. From *B. kolbi*, *B. rexroadi* differs in: over-all size of mandibular ramus, incisors, and molars smaller; anterior median fold of m1 present, though poorly developed.

From *B. brachygnathus*, *B. rexroadi* differs in: over-all size of mandibular ramus smaller; m3 larger; posterior cusps (hypoconid and entoconid) elongated; diastema shorter, less acutely recurved; incisors less proodont; cingular ridges of m1 and m2 less well-developed.

From *B. musculus*, *B. rexroadi* differs in: over-all size of mandibular ramus less; cingular ridges of m1 and m2 less well-developed; incisors smaller, more proodont; molars less depressed.

From *B. taylori*, *B. rexroadi* differs in: m3 more triangular, posterior part narrower; mental foramen closer to anterior root of m1; masseteric ridge closer to alveolus of m1; incisor shorter, more proodont; molars more depressed.

Remarks.--Two maxillary tooth-rows and associated parts were studied. On one of these specimens, the M2 has a well-developed mesostyle; the anterior median fold of M1 is also well-developed. The other specimen possesses a low cingular ridge (enteroloph) between the protocone and the hypocone, a reduced cingular ridge (mesoloph) between the paracone and metacone of M1. On the second molar, M2, a mesostyle joins with the mesoloph somewhat in the fashion indicated by Hooper (1957:9, encircled number 2).

Baiomys kolbi Hibbard

Baiomys kolbi Hibbard, Trans. Kansas Acad. Sci., 55:201, June 18, 1952; Hibbard, Papers Mich. Acad. Sci., Arts and Letters, 38:403, April 27, 1953.

Type.--No. 24846, Univ. Michigan; right mandibular ramus bearing m1-m3 and incisor; Fox Canyon, upper Pliocene, Rexroad formation, Rexroad fauna, Univ. Michigan Locality K1-47, sec. 35, T. 34 S, R. 30 W, XI Ranch, Meade County, Kansas.

Referred material.--Univ. Michigan Nos. 24845-24848, 27494, 27497, 27499, 28566, 28861, 28878, 28880-28882, 28884, 28886.

Diagnosis.--Ramus of medium size to large for the genus; lower incisor short, narrow transversely, proodont; anterior median fold of m1 reduced or absent; cingular ridges of m1 and m2 moderately well-developed; m3 large relative to m1 and m2; average and extreme measurements of lower molars of seven specimens are, 3.0 (3.0-3.1).

Comparisons.--For comparisons with *B. sawrockensis* and *B. rexroadi*, see accounts of those species. From *B. brachygnathus*, *B. kolbi* differs in: molar row longer; m3 and jaw larger; diastema longer; masseteric ridge not so far forward; molars more depressed.

From *B. minimus*, *B. kolbi* differs in: molar row longer; m3 larger; jaw larger; diastema not so acutely curved; incisor shorter, narrower transversely, more proodont.

From *B. musculus*, *B. kolbi* differs in: anterior median fold of m1 slightly developed or absent, instead of well-developed; m3 larger (not reduced), external reentrant valley broad and extending farther across crown of tooth; incisor smaller, and more proodont; cingular ridges of m1 and m2 less well-developed.

From *B. taylori*, *B. kolbi* differs in: molars larger, more depressed; incisor shorter, more proodont; m3 smaller relative to m1 and m2; external reentrant valley of m3 broad, extending farther across crown of tooth.

Remarks.--The slight development or absence of the anterior median fold in *kolbi* suggests that it was specialized. The anterior median fold is well-developed in all species of *Baiomys* save *B. brachygnathus* and *B. taylori*, in which the fold is only slightly developed or absent. *B. kolbi* may have paralleled *B. taylori* in specialization for a diet of grasses and for a life in open country.

Baiomys brachygnathus (Gidley)

Peromyscus brachygnathus Gidley, U. S. Geol. Surv. Prof. Papers, 131:124, March 15, 1922.

Baiomys brachygnathus, Hibbard, Amer. Midland Nat., 26:352, September, 1941.

P. [eromyscus] brachygnathus, Wilson, Carnegie Inst. Washington Publ., 473:33, May 21, 1936.

Type.--No. 10501, U. S. Nat. Mus.; right mandibular ramus bearing m1-m3, and incisor; 2 mi. NE Curtis Ranch house, near a line between sec. 28 and 29, T. 18 S, R. 21 E, Mid-Pleistocene (Hibbard, 1958:25), Cochise County, Arizona.

[Pg 591]

Referred material.--None.

Diagnosis.--Ramus small for the genus; m3 reduced; jaw reduced anteroposteriorly; incisor short, slender, proodont; cingular ridges well-developed, posterior ectolophid continuous from protoconid to hypoconid in m1 and m2; diastema short; length of molar row 2.8 mm.

Comparisons.--For comparisons with *B. rexroadi* and *B. kolbi,* see accounts of those species. From *B. minimus, B. brachygnathus* differs in: jaw not so slender anteriorly; masseteric ridge not so far anterior; cheek-teeth slightly broader, less depressed, therefore, more hypsodont; incisor shorter, more proodont.

From *B. sawrockensis, B. brachygnathus* differs in: molar row slightly longer; teeth slightly less depressed; masseteric ridge extends farther anteriorly; incisors more proodont.

From *B. musculus, B. brachygnathus* differs in: jaw smaller; molar row slightly shorter; molars less depressed; incisors slender, shorter, narrower, and more proodont.

From *B. taylori, B. brachygnathus* differs in: incisor more slender, shorter, more proodont; diastema shorter.

Remarks.--The molar teeth of *B. brachygnathus,* although worn, resemble those of *B. taylori* more than those of any known fossil species. Gidley (1922:124) stated that the absence of the divided anterior lobe of the first molar (anterior median fold) in *brachygnathus* was one of the chief characters separating *brachygnathus* from *taylori*. In *taylori,* the anterior median fold characteristically is only slightly developed, and in some specimens is absent. *B. brachygnathus* differs from *taylori* chiefly in proodont incisors, which feature seems to preclude *brachygnathus* being ancestral to *taylori*. *B. brachygnathus* may have been a specialized divergence from *B. minimus.*

Baiomys minimus (Gidley)

Peromyscus minimus Gidley, U. S. Geol. Surv. Prof. Papers, 131:124, March 15, 1922.

Baiomys minimus, Hibbard, Amer. Midland Nat., 26:352, September, 1941; Gazin, Prof. U. S. Nat. Mus., 92(3155):488, 1942.

P. [eromyscus] minimus, Wilson, Carnegie Inst. Washington Publ., 473:33, May 21, 1936.

Type.--No. 10500, U. S. Nat. Mus.; left mandibular ramus bearing m1-m3 and incisor; 2 mi. S Benson, sec. 22, T. 17 S, R. 20 E, Late Pliocene (Blancan, Gazin, 1942:482), Cochise County, Arizona.

Referred material.--None.

Diagnosis.--Ramus small for the genus; molar teeth depressed; cingular ridges (ectolophids) of m1 and m2 well-developed; anterior median fold present (appearing larger owing to chip of enamel missing); external reentrant fold of m3 progresses half way across crown of tooth; diastema short; incisor moderately large, recurved; length of molar row, 2.6 mm.

Comparisons.--For comparisons with *B. brachygnathus, B. kolbi,* and *B. sawrockensis,* see accounts of those species. From *B. rexroadi, B. minimus* [Pg 592] differs in: anterior median fold deeper; incisor longer, more recurved, less proodont; molars slightly more depressed (though worn).

From *B. musculus*, *B. minimus* differs in: over-all size of jaw and molars smaller; incisors shorter; masseteric ridge more depressed.

From *B. taylori*, *B. minimus* differs in: anterior median fold slightly deeper; molar teeth more depressed; cingular ridges on m1 and m2 better developed; masseteric ridge more depressed.

Remarks.--Gidley (1922:124) stated that *B. minimus* differed considerably from *B. taylori* in that the coronoid portion of the ascending ramus diverges at a wider angle from the alveolar part of the jaw. Study of large samples of lower jaws of *B. taylori* reveals considerable individual variation in the angle formed between the coronoid part of the jaw and the alveolar part.

B. minimus, except for its small size, is like *B. musculus* and is considered to be ancestral to that species.

PHYLETIC TRENDS

It seems that the important trends in phyletic development in the pygmy mice have been from an ancestral stock (see Figure 3) that possessed relatively brachydont teeth having raised cingular ridges (ectolophids and mesolophids) and relatively short orthodont to proodont incisors, to species having teeth more hypsodont on which cingular ridges were reduced, stylids were isolated or completely absent, and incisors were longer and more recurved or retrodont. *Baiomys sawrockensis*, or an unknown stock resembling it, might have been ancestral to the other known species. Of the four remaining fossil species, *B. kolbi* seems least likely to have been ancestral to the two living species, owing to its proodont incisors, reduction of cingular ridges, loss of an anterior median fold in m1, and long mandibular tooth-row. *B. kolbi* may have been an early, specialized derivation from the ancestral stock. From his knowledge of the habitats of *B. musculus*, the larger species, and *B. taylori*, the smaller species, Hibbard (1952:203) suggests that *B. kolbi*, a large species, might have inhabited lowlands, and *B. rexroadi*, a small species, highlands. I have no evidence to dispute this suggestion except that *B. musculus* has more prominent cingular ridges (or at least vestiges of this lophid condition) than either *B. kolbi* or *B. rexroadi*. *B. musculus* (see page 610) is less of an open grassland inhabitant than is *B. taylori*. Therefore, both *B. kolbi* and *B. rexroadi*, because of their poorly developed cingular ridges, might be expected to have lived in a relatively open grassland habitat.

The relationship of *B. rexroadi* to fossil species other than *B. kolbi* is not clear. Superficially, the former resembles *B. taylori*, but, owing to the specialized development of the molars of *rexroadi*, it could hardly have been ancestral to either of the living species. The resemblance of *B. rexroadi* to *B. taylori* may result from each having occupied the same ecological niche in different periods. The incisors of *B. rexroadi*, however, are much shorter than those of *B. taylori* and suggest somewhat different food habits.

B. minimus seemingly is more closely related to *B. sawrockensis* and *B. musculus* than to the other described species. The development of the cingular ridges leads one to suspect that *B. minimus* was the ancestor of *B. musculus*. *B. minimus* may have been derived from a *sawrockensis*-like stock and probably gave rise to *B. musculus*.

Hershkovitz (1955:643-644) suggests that "... primitive brachydont, buno-mesolophodont cricetines have survived ... in forested parts of the range," whereas "... the progressive branch of cricetines with mesoloph absent or vestigal, has become increasingly specialized for life in open country and a diet of grasses." Species of the genus *Baiomys* can be divided into two morphological groups. One group, composed of *B. sawrockensis*, *B. minimus*, and *B. musculus*, includes those species, the teeth of which were relatively brachydont and had prominently developed cingular ridges (ectolophids or mesolophids) or, at least, showed some development of these ridges. *B. sawrockensis* probably lived in semi-wooded to shrubby habitats. According to Hibbard (1953:409), "The Saw Rock Canyon fauna lived in that area at a time when conditions were comparable to the conditions at the time the Rexroad fauna lived." The conditions in which the Rexroad fauna lived are discussed by Hibbard (1941:95). Presumably, there were at least some well-wooded situations, and the climate was warm. *B. sawrockensis* probably inhabited denser vegetation than did *B. minimus* or than does *B. musculus*. The teeth of the second group (*B. kolbi*, *B. rexroadi*, *B. brachygnathus*, and *B. taylori*) lack cingular ridges or have them much reduced and have more hypsodont molars. The three fossil species probably inhabited relatively open grassland. This assumption is based largely on the known habitat of *B. taylori* (see page 632).

The suggested grouping, based on supposed similarities in niches inhabited by the extinct species, does not necessarily indicate degree of relationship. *B. taylori* probably was not derived from an ancestor like *B. rexroadi* or *B. kolbi*, although, in certain characters, the three species resemble one another. *B. kolbi* and *B. rexroadi* were already specialized in Blancan times, probably for living on grassland. *B. taylori* shows only a slight advance in specialization of molar structures compared to either of the aforementioned species but is slightly smaller and does have longer and more recurved incisors. If only morphological criteria of lower jaws were considered, without recourse to other data derived from the study of many samples of populations of the living species, time alone might account for the differences among *B. taylori*, *B. rexroadi*, and *B. kolbi*. The available evidence (see page 658) suggests, however, that *B. taylori* was derived from the *B. sawrockensis*-*B. minimus*-*B. musculus* line.

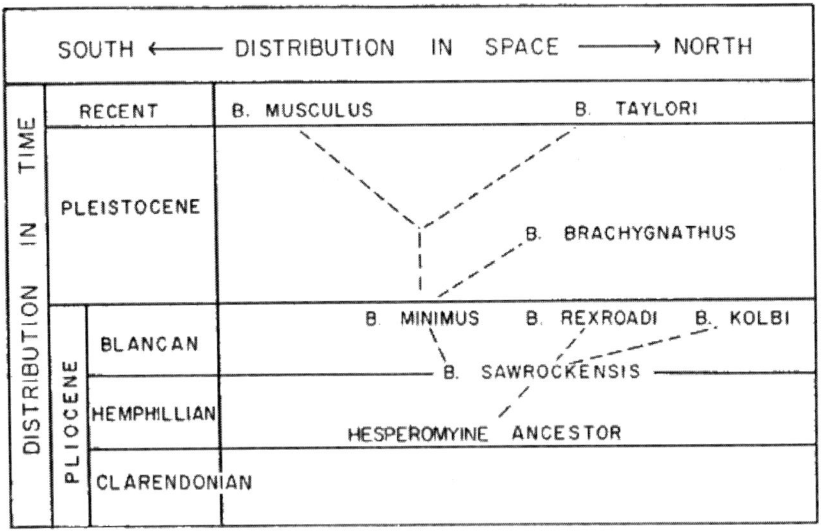

FIG. 3. Diagram indicating probable relationships of living and extinct species of pygmy mice.

Baiomys seems to have undergone little basic evolutionary and morphological change since Late Pliocene time. According to Simpson (1945:207), hesperomine rodents as a group have undergone little basic evolution, and "The rapid evolution of new genera was more a matter of segregation of characters in a group with a great variation than of the origin of significantly new characters." Perhaps, the living southern pygmy mouse retains many basic characteristics of one of the early North American cricetine-like stocks that emigrated to South America near the end of the Pliocene epoch. There is much to suggest close relationship of the pygmy mice to certain species of South American hesperomine rodents of the genus *Calomys*.

NON-GEOGRAPHIC VARIATION

Non-geographic variation in pygmy mice (variation in a single population resulting from age, individual, seasonal, and secondary sexual differences) has been but little studied in the past. Mearns (1907:381) figured progressive stages of wear on the teeth of *B. taylori*; Osgood (1909:252) and Blair (1941:380) referred to changes in dentition, weights, and pelages.

The largest samples available for this study were 47 *B. taylori* from the vicinity of Altamira (6 mi. N, 6 mi. W; 5 mi. N, 5 mi. W; 1 mi. S),

Tamaulipas, and 44 *B. musculus* from El Salvador (1 mi. S Los Planes, and 1 mi. NW San Salvador--two localities 3 miles apart).

VARIATION WITH AGE

Specimens of both species were segregated into five categories: Juveniles, young, subadults, adults, and old adults. Juvenal and young pygmy mice are readily separable from the other three categories; subadults are less easily distinguished from adults. In order to obtain an accurate understanding of geographic variation in these mice, only adults should be used in making taxonomic comparisons.

Juveniles.--Nestling mice yet unweaned; sutures in cranium incompletely closed; bony parts of skull fragile; M3 and m3 not erupted or only partly erupted and not protruding above margins of alveoli.

At birth, juveniles are pink, without pelage except for the mystacial vibrissae and a few hairs about the eye. Blair (*op. cit.*:381) recorded changes with age in color of the skin of new-born and suckling pygmy mice. Data obtained by me from three litters born in captivity agree with his findings. Pygmy mice are weaned when 17 to 24 days old. At that time, the mice possess a fine, but not dense, dusky-gray fur.

Young.--Weaned mice; cranium fragile; sutures between frontals and parietals, interparietal and parietals, basioccipital and basisphenoid, basisphenoid and presphenoid, premaxillaries and maxillaries widely open; M3 and m3 erupted beyond margins of their alveoli (molars erupt from anterior to posterior; M3 and m3, therefore, are last to erupt); in some specimens, molars slightly worn; pelage still dusky and relatively fine and sparse.

Subadults.--Sutures between bones of skull less widely open than in young; epiphyses of long bones incompletely coalesced to shaft; relative to length of skull, braincase higher and rostrum shorter than in adults; all cusps worn, but dentine not occlusally confluent; primary first and second folds of third upper molars present; primary first fold and major fold of lower molars visible; pelage a subtle mixture of colors of young and adult, but resembling most that of adult; molts into postjuvenal pelage between 46 and 50 days.

Adults.--Sutures of skull, and those between epiphyses and shaft of long bones obliterated except that, in some mice, sutures of skull persist between frontoparietal, and interparietal; cusps of molars so worn that dentine occlusally confluent; small island of enamel in third upper and lower molars of some specimens; relative to length of skull, cranium lower, rostrum longer, and interorbital region narrower than in subadult; cranium appears to be more flattened dorsoventrally; between subadult and adult stages, principal growth occurs in basioccipital, basisphenoid, frontals, and parietals; nasals grow less.

Although all bones of the skull grow in the subadult and early adult stages (see table 1), the above-named bones grow faster than others and thus cause the general flattening of the skull, typical of adults (similar to that reported by Hoffmeister, 1951:7). The body continues to lengthen, accounting for the increase in total length of the adult (see table 1). Hind foot, tail and ear, reach their maximum lengths by subadult stage. Adult pelage has been acquired, and the color is brighter than in either subadults or old adults.

Old Adults.--Characterized principally by well-worn molars; only thin peripheral band of enamel along with slight evidence of any primary or secondary folds on any teeth remain; all bones of skull coalesced; epiphyses and shafts of long bones ankylosed; small bony protuberances on many skulls; pelage usually ragged, tips of the hairs being worn away; white flecking and spotting not common, but occurs in some adults.

TABLE 1.--AVERAGE AND EXTREME MEASUREMENTS (IN MILLIMETERS) OF SKULLS OF FIVE AGE-GROUPS OF BAIOMYS TAYLORI FROM VIC. (SEE P. 595) ALTAMIRA, TAMAULIPAS, MEXICO.

Age groups	Number examined
Juvenile	3
Young	3
Subadult	14
Adult	19
Old adult	8

Total length

77.0
(74-79)
92.6
(89-96)
97.6
(91-103)
99.9
(93-105)
101.6
(98-107)

Length of tail

27.3
(24-29)
39.3
(37-41)
40.4
(36-43)
39.8
(35-45)
40.9
(38-45)

Length of body

49.6
(49-50)
53.3
(52-55)
57.0
(51-61)
60.0
(56-67)
60.7
(57-67)

Length of hind foot
11.0
(11)
13.6
(13-14)
14.3
(13.5-15.0)
14.5
(14-15)
14.2
(13-15)

Occipitonasal length
14.2
(13.6-15.2)
16.3
(15.8-16.9)
17.1
(16.7-17.6)

17.7
(17.2-18.3)
17.8
(17.6-18.1)

Zygomatic breadth
8.1
(7.8-8.6)
8.7
(8.6-8.8)
8.9
(8.6-9.3)
9.3
(9.0-9.6)
9.4
(9.1-9.6)

Interorbital breadth
3.4
(3.3-3.5)
3.4
(3.3-3.6)
3.4
(3.3-3.6)
3.6
(3.4-3.8)
3.5
(3.3-3.6)

Incisive foramina (length)
2.9
(2.8-2.9)
3.5
(3.4-3.6)
3.7
(3.6-3.9)
3.9
(3.6-4.1)
3.9
(3.5-4.0)

Depth of cranium
5.9
(5.6-6.2)
6.5
(6.3-6.8)
6.5
(6.2-6.8)
6.7
(6.4-7.0)
6.8
(6.5-7.1)

Alveolar length, upper molars

2.7
(2.5-2.8)
2.9
(2.9-3.0)
2.9
(2.8-3.1)
3.0
(2.9-3.2)
3.0
(3.0-3.1)
Postpalatal length
4.8
(4.5-5.3)
5.9
(5.8-6.0)
6.2
(5.8-6.6)
6.5
(6.2-7.2)
6.5
(6.3-6.7)
Breadth of braincase
8.1
(7.8-8.7)
8.5
(8.5)

8.4
(8.0-8.7)
8.6
(8.3-8.9)
8.6
(8.4-8.8)

SECONDARY SEXUAL VARIATION

The method employed by Dice and Leraas (1936:2) was used to measure the secondary sexual differences, if there were any, in each of several age classes. As pointed out by Hooper (1952b:11), individual variation in small samples can obscure secondary sexual differences. The samples of *B. taylori* from the vicinity (see page 595) of Altamira, Tamaulipas, and the samples of *B. musculus* from El Salvador (table 2) were large enough to prevent individual variation from obscuring sexual differences. Nevertheless, no significant secondary sexual differences were found in either *B. taylori* or *B. musculus* (see table 2). Therefore, the sexes have been considered together for purposes of geographic studies.

TABLE 2.--ANALYSIS OF SECONDARY SEXUAL VARIATION IN ADULT B. TAYLORI VICINITY OF (SEE P. 595) ALTAMIRA, TAMAULIPAS, AND ADULT B. MUSCULUS FROM EL SALVADOR (SEE P. 595).

(ONE STANDARD DEVIATION ON EITHER SIDE OF THE MEAN IS GIVEN.)

Character
Baiomys taylori
Baiomys musculus
21 Males
18 Females
17 Males
13 Females

Total length
98.4 +- 2.95

100.5 +- 4.72
112.04 +- 5.49
113.12 +- 4.23

Length of tail
40.1 +- 2.31
40.3 +- 2.39
47.12 +- 2.95
45.70 +- 2.92

Length of body
57.83 +- 1.65
60.10 +- 4.13
66.67 +- 3.97
67.75 +- 2.38

Length of
hind foot
14.21 +- .53
14.44 +- .51
15.60 +- .49
15.38 +- .64

Length of ear

10.00 +- .00
10.00 +- .00
11.80 +- .65
12.00 +- .41

Occipitonasal length
17.48 +- .40
17.47 +- .47
19.32 +- .35
19.04 +- .44

Zygomatic breadth
9.17 +- .33
9.15 +- .30
9.84 +- .21
9.91 +- .28

Least interorbital breadth
3.53 +- .11
3.48 +- .11
3.88 +- .08
3.88 +- .12

Postpalatal length

6.35 +- .19
6.38 +- .30
7.11 +- .15
6.95 +- .20

Depth of cranium

6.65 +- .24
6.61 +- .17
7.10 +- .18
7.08 +- .18

Incisive foramina (length)

3.82 +- .15
3.81 +- .18
4.43 +- .11
4.35 +- .14

Length of rostrum

5.87 +- .20

5.88 +- .21
6.81 +- .16
6.66 +- .31

Breadth of braincase

8.54 +- .23
8.52 +- .12
9.84 +- .38
9.52 +- .20

Alveolar length, upper molars

2.98 +- .08
3.01 +- .08
3.20 +- .09
3.24 +- .10

INDIVIDUAL VARIATION

Length of tail varied more than any other measurement used by me in taxonomic comparisons. Clark (1941:298), Hoffmeister (1951:16), and Van Gelder (1959:239) point out that external measurements generally are more variable than measurements of the cranium, probably because different techniques of measuring are employed by different collectors. As can be noted in table 3, females varied more than males.

In the 3520 specimens examined, an extra tooth was observed in only one (see Hooper, 1955:298). The left mandibular tooth-row of an adult male (USNM 71539) from Omentepec, Guerrero, is worn more than the right one. Irregularities in number of teeth and abnormalities in individual teeth seem to be rare in pygmy mice.

TABLE 3.--INDIVIDUAL VARIATION: COEFFICIENTS OF VARIATION FOR DIMENSIONS OF EXTERNAL AND CRANIAL PARTS IN A POPULATION OF B. MUSCULUS AND B. TAYLORI.

Measurement
Baiomys taylori
Baiomys musculus
Vic. (see page 595)
Altamira, Tamaulipas
Vic. (see page 595)
El Salvador
21 Males
C. V.
18 Females
C. V.
17 Males
C. V.
13 Females
C. V.

Total length

3.0

4.7

4.9

3.7

Length of tail

5.7

5.9

6.2

6.4

Length of body

2.8

5.0

5.9

3.5

Length of hind foot

3.7

3.4

3.0

4.1

Length of ear

0.0

0.0

5.5

3.3

Occipitonasal length

2.2

2.7

1.8

2.3

Zygomatic breadth

3.6

3.3

2.2

2.7

Interorbital breadth

3.2

3.3

2.2

2.9

Incisive foramina (length)

3.8

4.6

2.5

3.2

Depth of cranium

3.6

2.5

2.5
2.5
Alveolar length,
upper molars
2.7
2.5
2.8
3.2
Postpalatal length
3.1
4.7
2.1
2.9
Length of rostrum
3.3
3.6
2.4
4.7
Breadth of braincase
2.7
1.4
4.0
4.9

The posterior margin of the bony palate varies from semicircular to nearly V-shaped. The suture between the nasals and frontals varies from V-shaped to truncate to W-shaped. The maxillary part of the zygoma varies from broad to slender in dorsoventral width in both species.

PELAGE AND MOLTS

There are three distinct pelages, juvenal, postjuvenal, and adult. The sequences of molt and change of pelage from the juvenal, to the postjuvenal, and from it to adult, are essentially as reported for *Peromyscus* by Collins (1918:78-81; 1924:58-60) and Hoffmeister (1951:5). The juvenal pelage is uniformly dusky gray throughout except for the paler gray on the venter. In most juvenal mice, the yellow to ochraceous pigments of the subterminal bands are reduced or absent. Unlike *Peromyscus*, *Baiomys* has bright brownish hairs on the head as the first evidence of the postjuvenal molt (see Figure 4, part a). Blair (1941:381) reports adult pelage in pygmy mice being evident first at an age of 46 days. Two of my juveniles born in captivity began the postjuvenal molt on the 38th and 40th days. The area of new hairs on the head spreads most rapidly posteriorly. New hair appears ventrally and laterally at the end of 46 days (see Figure 4, part b). Hair replacement proceeds more slowly after the "saddle back" stage (described in *Peromyscus* by Collins, 1918:80) has been reached. That stage was reached in two pygmy mice at 52 days (see Figure 4, part c). Areas immediately posterior to the ears, in the scapular region, molt last. The postjuvenal pelage was seemingly complete in one captive pygmy mouse at the end of 60 days. Another captive failed to complete its growth of new pelage until two additional weeks had elapsed. Length of time required to molt in pygmy mice is about the same as that reported by Layne (1959:72) in *Reithrodontomys*.

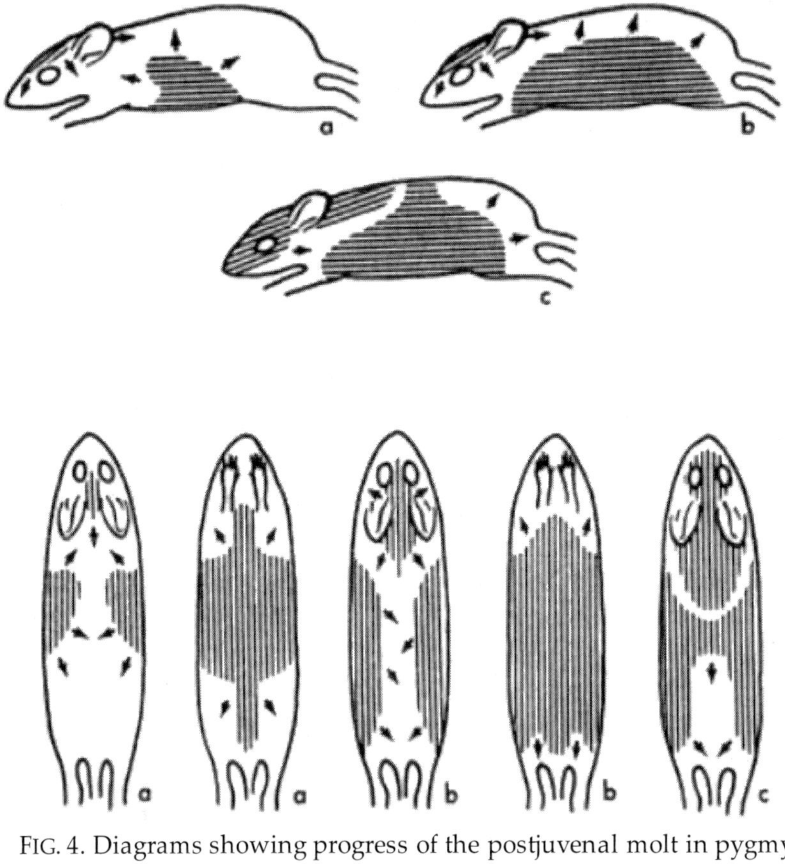

FIG. 4. Diagrams showing progress of the postjuvenal molt in pygmy mice.
For explanation of a, b, and c, see text. All approximately $^2/_3$ natural size.

If, after the postjuvenal molt, a distinct adult pelage is acquired it is difficult to separate it from the annual replacement of pelage in adults at the beginning of the rainy season. Adults of both species have been found in molt in all months of the year. To the north, in Texas, the pelage of winter-taken specimens is denser and slightly more reddish than that of specimens taken in spring and summer. In the two last mentioned seasons, the pelage is more uniformly gray. To the south, in Mexico, the pelage is heavy and long in most specimens taken in the rainy season. The percentage of specimens in molt immediately before the rainy season and immediately before the dry season is slightly higher than in specimens taken at other times of the year. The adult or seasonal molt (both loss of old pelage and growth of new) resembles that in *Peromyscus truei gilberti*, described by Hoffmeister (1951:6) as proceeding "posteriorly as a wave over the entire back." The new hair is slightly brighter than the old. Old adults are usually in ragged pelage regardless of season; possibly only one regular annual change of pelage occurs in most animals before they die. Only one case of melanism was observed among all the specimens of both species examined. It was a young male *B. t. taylori*, KU 35943, from 6 mi. SW San Geronimo, Coahuila, possessing black hairs throughout. Its hairs are longer and finer than those on specimens of comparable age and sex. No albino was found, although Stickel and Stickel (1949:145) record one--an adult male of *B. taylori*.

TAXONOMIC CHARACTERS AND RELATIONSHIPS

External parts.--Length of body, foot, ear, and tail are useful when considered together in distinguishing species and subspecies. I found as Hooper (1952a:91) did that length of ear in combination with length of hind foot suffices to identify nearly all specimens to species, especially where the two species occur together.

Pelage.--Color in adults is of especial value in subspecific determination; the manner in which it varies geographically is described on pages 609, 630.

Skull.--Difference in occipitonasal length and zygomatic breadth, both having low coefficients of variation, are useful in separating species, especially where they are sympatric. Shape of presphenoid, nasals, interparietal, frontoparietal sutures, and length and degree of the openings of the incisive foramina are useful in delimiting subspecies. The rostrum of *B. taylori*, in front of the frontonasal suture, is deflected three to five degrees ventrally in 85 per cent of the adults examined, and in *B. musculus* is less, or not at all, deflected.

Teeth.--Alveolar length of the upper and lower molar tooth-rows aids in distinguishing fossil and Recent species, and to a lesser degree in delimiting subspecies. Occlusal pattern is useful in estimating the relationship of fossil and living species. Degree of development of the mesostyle, mesostylid, mesoloph, and mesolophid have been useful in determining relationship between fossil and living species as well as useful in separating the living species. Rinker (1954:119) and Hooper (1957:48) have shown the degree of variation in dental patterns in *Peromyscus*, *Sigmodon*, and *Oryzomys*, mice thought to be closely related to *Baiomys*. In pygmy mice, however, the dental patterns are relatively constant. The lophs and styles are subject to some geographic variation but, nevertheless, are useful in estimating relationships.

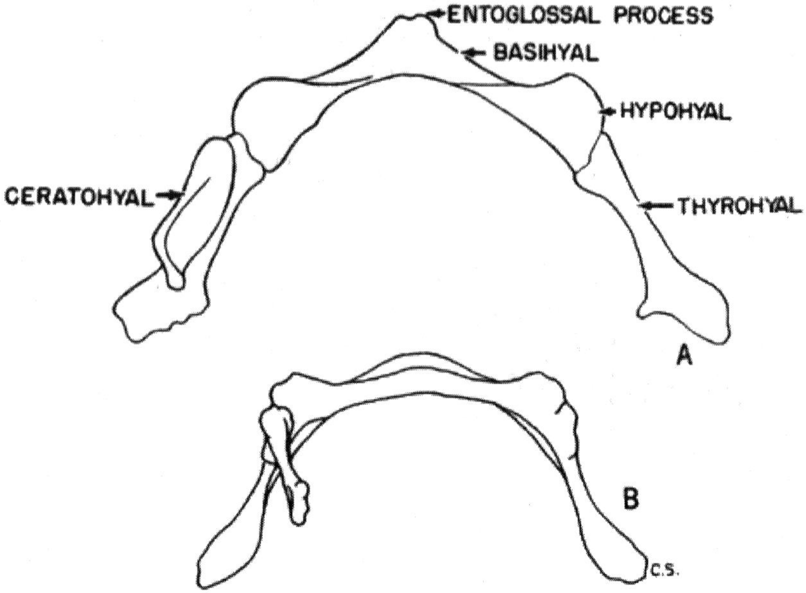

FIG. 5. Ventral view of hyoid bones. x 18.

A. *Baiomys musculus brunneus,* adult, female, No. 30182 KU, Potrero Viejo, 1700 feet, Veracruz.

B. *Baiomys taylori analogous,* adult, female, No. 36761 KU, 2 mi. N Ciudad Guzman, 5000 feet, Jalisco.

Hyoid apparatus.--Shape and, to a lesser extent, size of the hyoid apparatus differentiate nearly all specimens of *B. taylori* from all those of *B. musculus*. The hyoid of *B. taylori* differs from that of *B. musculus* principally in the shape of the basihyal. It possesses an anteriorly pointed entoglossal process in *B. musculus,* and is not rounded to completely absent as in *B. taylori* (see Figure 5). The shoulders of the basihyal protrude anteriorly in *B. musculus,* and are not flattened as in *B. taylori*. The total length was measured in a sample of 55 basihyals of *B. musculus,* and was compared to the total length of a sample of 80 basihyals of *B. taylori*. The means of the two samples differ significantly at the 95 per cent level; the mean plus two standard errors of *B. musculus* and *B. taylori,* are, respectively, 2.43 +- .02; 2.18 +- .03. There is sufficient overlap of the samples (mean plus one standard deviation of *B. musculus* and *B. taylori,* respectively: 2.43 +- .15; 2.18 +- .15) to make the total length of the basihyal of only secondary importance in distinguishing species, but shape and total length of the basihyal, when considered together, serve to identify all specimens to species. When length of the basihyal is plotted against occipitonasal length (see Figure 6), all specimens studied, regardless of age or geographical origin, were separated at the level of species. The hypohyals of *B. taylori* seemingly remain distinct throughout life; those of *B. musculus* completely fuse in some adults. The ceratohyals are highly variable in shape and of little taxonomic use.

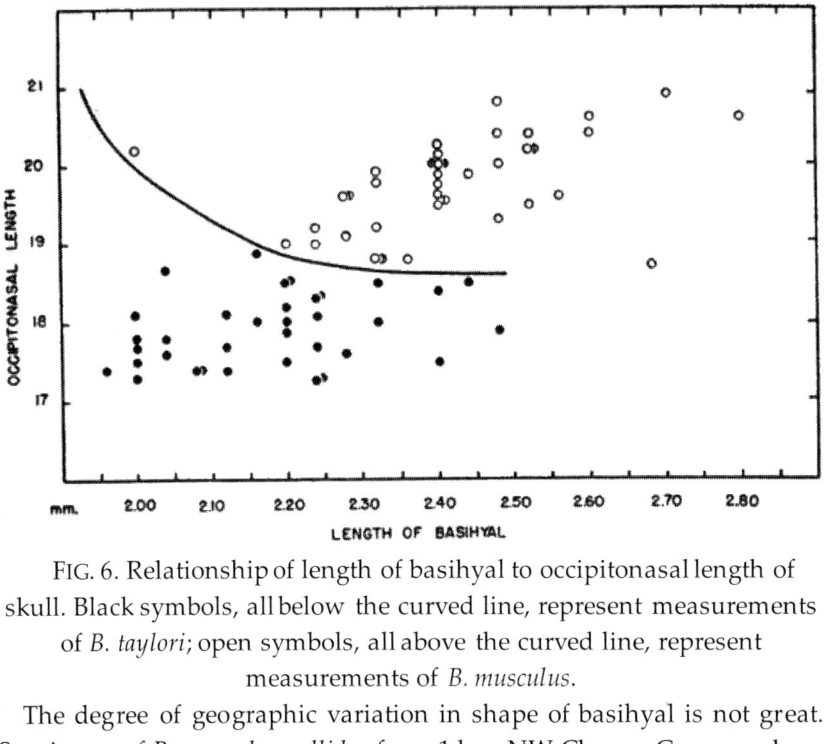

FIG. 6. Relationship of length of basihyal to occipitonasal length of skull. Black symbols, all below the curved line, represent measurements of *B. taylori*; open symbols, all above the curved line, represent measurements of *B. musculus*.

The degree of geographic variation in shape of basihyal is not great. Specimens of *B. musculus pallidus* from 1 km. NW Chapa, Guerrero, have a small indentation on the anteriormost part of the entoglossal process. The shoulder of the basihyal is directed less forward in specimens of *B. taylori taylori* from 6 mi. N, 6 mi. W Altamira, Tamaulipas, than in other specimens of the species. The variations observed seemed not to be clinal.

According to White (1953:548) the hyoid, like the baculum (Burt, 1936:146), is little influenced by changes in external environment and may serve to clarify intergeneric relationships. Hyoids of both species of *Baiomys* are smaller than hyoids of all subgenera of *Peromyscus*. In shape, the hyoids of *Baiomys* resemble those of *Ochrotomys nuttalli* (as explained on page 605, *Ochrotomys* is here accorded generic, instead of subgeneric, rank). In size, the hyoid of both species of *Baiomys* resembles that in *Reithrodontomys*. Sprague (1941:304) reports a resemblance in shape between the ceratohyals of *Baiomys* and *Reithrodontomys*. The thyrohyals differ from those of *Reithrodontomys*, being less boot-shaped, and having a slight terminal expansion as in *Ochrotomys* (see Sprague, *loc. cit.*). In shape, the large basihyal of *Onychomys* resembles the smaller one of *B. musculus*. The basihyal of *Oryzomys* lacks the entoglossal process present in *Baiomys*. On the basis of shape of hyoid, *Baiomys* seems to be most closely related to *Ochrotomys*.

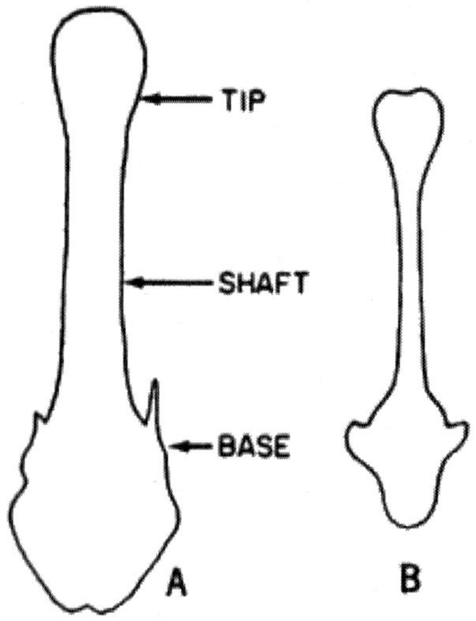

FIG. 7. Dorsal view of bacula. x 16.

A. *B. musculus brunneus*, adult, No. 24336 KU, 3 kms. W Boca del Rio, 10 feet, Veracruz.

B. *B. taylori taylori*, adult, No. 35937 KU, 6 mi. SW San Geronimo, Coahuila.

Baculum.--Of *Baiomys*, 166 bacula were processed, using the method of White (1951:125), and studied. They provide characters of taxonomic worth at the level of species and aid in evaluating generic relationships.

The baculum of *B. taylori* differs from that of *B. musculus* in: shaft narrow; wings anterior to base projecting dorsolaterally instead of anteriorly; anterior part knob-shaped having indentation at tip, instead of anterior part spatulate-shaped (in some) to knob-shaped (see Figure 7), without indentation; significantly shorter (see Table 4).

TABLE 4.--LENGTH OF BACULA

	Number of specimens	Average length	3 x standard	1 standard	Range
B. taylori	108	2.535	.078	.274	2.00-3.12
B. musculus	58	3.324	.090	.233	2.80-3.88

In each of the two species, individual and geographic variation in the baculum is slight; its length varies insignificantly according to age. Excluding juveniles contained in Table 4, but including young and subadults, only three bacula of *B. taylori* were longer than 3 mm., and only one baculum of *B. musculus* (a young) was shorter than 3 mm. The total length of the baculum, considered together with its shape, serves to identify to species all specimens examined by me.

The bacula of both species of *Baiomys* were compared with bacula of *Akodon, Scotinomys, Holochilus, Oryzomys, Zygodontomys, Reithrodontomys, Thaptomys,* and *Calomys* and illustrations of bacula by Blair (1942:197, 200) of *Peromyscus* (subgenera *Peromyscus, Haplomylomys, Podomys*), *Ochrotomys*, and material at the University of Kansas Museum of Natural History of *Megadontomys*. Shape of baculum most resembled that of *Ochrotomys* and *Calomys*. The bacula of *Baiomys*, as pointed out by Blair (*op cit.*:203), differ as much from those of the genus *Peromyscus* as do the bacula of *Reithrodontomys* and *Onychomys*. In size of baculum, *Baiomys* resembles *Ochrotomys*. Blair (*op. cit.*:202) pointed out that the length of the baculum of *B. taylori subater* was contained in the length of the animal's body 20.3 times, and 24.2 times in the length of that of *Ochrotomys nuttalli*. The length of the baculum of *B. musculus* (average of 58 specimens without regard to subspecies) is contained in the length of the body (of specimens from which the bacula were removed) 22.7 times, a figure approaching that in *Ochrotomys*. When bacula of both species of *Baiomys* were compared to those of *O. nuttalli*, bacula of *B. musculus* were found to most closely resemble those of *O. nuttalli*. The baculum of a single specimen of *Calomys* (*C. laucha*) was contained in the length of the body 15.5 times. In general shape, as well as in possession of an anterior knob and the position of the expanded posterior wings, the baculum of *C. laucha* resembles the baculum of *Ochrotomys* and *Baiomys musculus*.

Blair (*op. cit.*:201) considers generic *versus* subgeneric rank for *Ochrotomys*, and on the basis of studies of the phallus Hooper (1958:23) stated that "it is clear that *nuttalli* should be removed from *Peromyscus* and should be listed as *Ochrotomys nuttalli* (Harlan)." I agree with Hooper (*loc. cit.*) and point out that on the basis of the baculum, there is less of a hiatus between *Baiomys* on the one hand, and *Ochrotomys* and *Calomys* on the other hand, than there is between any one of those three genera and *Peromyscus*.

White (1953:631) reported that the baculum of chipmunks might indicate relationships more clearly than do skulls and skins. He thought that skulls might more quickly than bacula reflect the habitus of the animal. The resemblance in cranial morphology between *Peromyscus* and *Baiomys* is judged to be the result of such a convergence of habitus and the baculum in *Baiomys* is thought to reflect relationships more accurately than does the skull.

Auditory ossicles.--Examination of a number of auditory ossicles of *Baiomys* reveals constant interspecific differences in the malleus and incus. There is only slight individual variation, slight variation with age, and no secondary sexual variation. In *Baiomys taylori* the orbicular apophysis of the malleus (see Figure 8, A) is rounded to nearly ovoid; the anterior process is pointed, and the neck is short, being slightly recurved. The body of the incus is round and the short process is elongate. The sides of the long limb of the incus are nearly parallel. The lenticular process is relatively large. The posterior and anterior crus of the stapes are bowed, and the muscular process is either absent or much reduced.

In *Baiomys musculus*, the orbicular apophysis of the malleus (see Figure 8, B) is round to oblong, and less ovoid than in *B. taylori*; the anterior process is less acutely pointed than in *B. taylori*, and the neck is long, less recurved than in *B. taylori*. The body of the incus, though tending to be round, is more flattened, and the short process is knob-shaped, not elongated. The sides of the long limb of the incus are not parallel. The lenticular process is, relative to the size of the incus, small. The posterior and anterior crus of the stapes are more nearly straight than in *taylori*. A prominent muscular process occurs on the posterior crus.

The auditory ossicles of representative species of all the subgenera of *Peromyscus* were studied as were the ossicles of *Onychomys, Ochrotomys, Oryzomys, Akodon, Thaptomys, Zygodontomys, Calomys, Reithrodontomys,* and *Holochilus*.

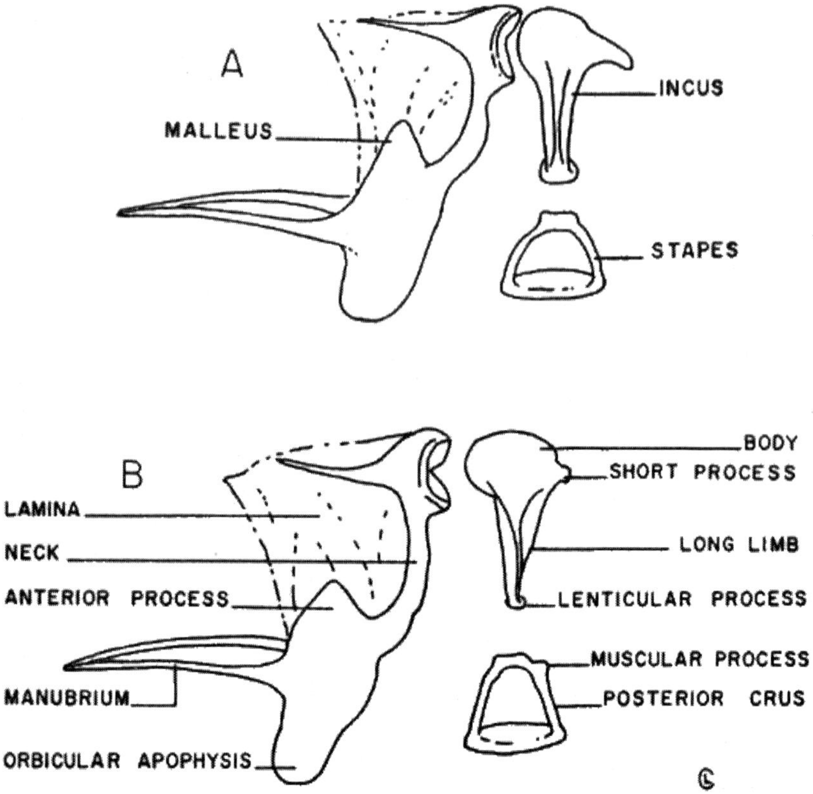

FIG. 8. Lateral views of auditory ossicles. x 20.

A. *B. taylori analogous*, adult, female, No. 28104 KU, 4 kms. ENE Tlalmanalco, 2290 meters, Estado de Mexico.

B. *B. musculus pallidus*, adult, male, No. 28346 KU, Cahuilotal, Sacacoyuca, 960 meters, Guerrero.

The general plan of structure of the auditory ossicles in *Baiomys* resembles that in *Calomys, Akodon,* and *Thaptomys*. The ossicles of *Calomys* and *Thaptomys*, in particular, closely resemble the auditory ossicles of *Baiomys musculus*. The short process of the incus is knoblike in *Calomys* and *Thaptomys*, and the general conformation of malleus and stapes in those two genera is nearly identical to that in *B. musculus*. In *Akodon*, the anterior and posterior crus of the stapes is more rounded than in *B. musculus*, resembling that in *B. taylori*.

Reithrodontomys differ from *Baiomys* in having a more elongate orbicular apophysis on the body of the malleus, an elongated short limb on the incus, and a stapes having anterior and posterior crura bowed as in mice of the genus *Peromyscus*.

In *Ochrotomys*, the orbicular apophysis of the malleus resembles the orbicular apophysis of *B. musculus*, but the short process of the incus is longer, resembling the short process of *B. taylori*. In general conformation of the malleus, incus, and stapes, *Ochrotomys* shows closer resemblance to *B. taylori* than to *B. musculus*.

In *Holochilus* the anterior crus and posterior crus of the stapes are similar to those in *B. musculus*, but in shape and size of malleus and incus, *Holochilus* differs considerably from *B. musculus* and *B. taylori*.

In *Zygodontomys*, size and shape of the ossicles differ greatly from those of *Baiomys*.

In the genus *Peromyscus*, only *Peromyscus floridanus* (subgenus *Podomys*) possesses a knoblike short process on the incus similar to that in *B. musculus*; representatives of the other subgenera examined possess an elongated short limb on the incus. The conformation of the ossicles of both *Onychomys* and *Oryzomys* appears to be more nearly like that in *Peromyscus* than that of *Baiomys*.

On the basis of shape and size of auditory ossicles, *Baiomys* resembles South American hesperomines (*Calomys* and *Thaptomys*) rather than North American hesperomines.

Genus Baiomys True

1894. *Baiomys* True, Proc. U. S. Nat. Mus., 16:758, February 7.

Type, *Hesperomys (Vesperimus) taylori* Thomas.

Diagnosis.--Size small (total length in adults, 93-135); tail shorter than head and body; hind foot in adults 12-17; ears small (8-12) and rounded; upper parts blackish sepia to ochraceous-buff; underparts slaty gray to white or pale buffy; eyes small; hind feet having six plantar pads, soles nearly naked except for some hairs on anterior parts of soles and anteriorly to base of toes and between toes; occipitonasal length of skull in adults, 17.0-21.5; zygomatic breadth, 9.0-11.5; coronoid process of mandible well developed, strongly recurved; ascending ramus of mandible short and erect; anterior palatine foramina (incisive foramina) long, usually terminating posterior to plane of the front of first molars; posterior palatine foramina nearly opposite middle of M2; interorbital space wide relative to widest part of frontals; nasals projecting only slightly over incisors; condyle terminal; upper incisors relatively heavy;

primary first fold of M3 obliterated at an early stage of wear; major cusps of upper and lower anteriormost two molars alternating, more so in m1-m2 than in M1-M2, dental formula I/i, 1/1; C/c, 0/0; P/p, M/m, 3/3 = 16.

For distribution of the genus, see Figure 9.

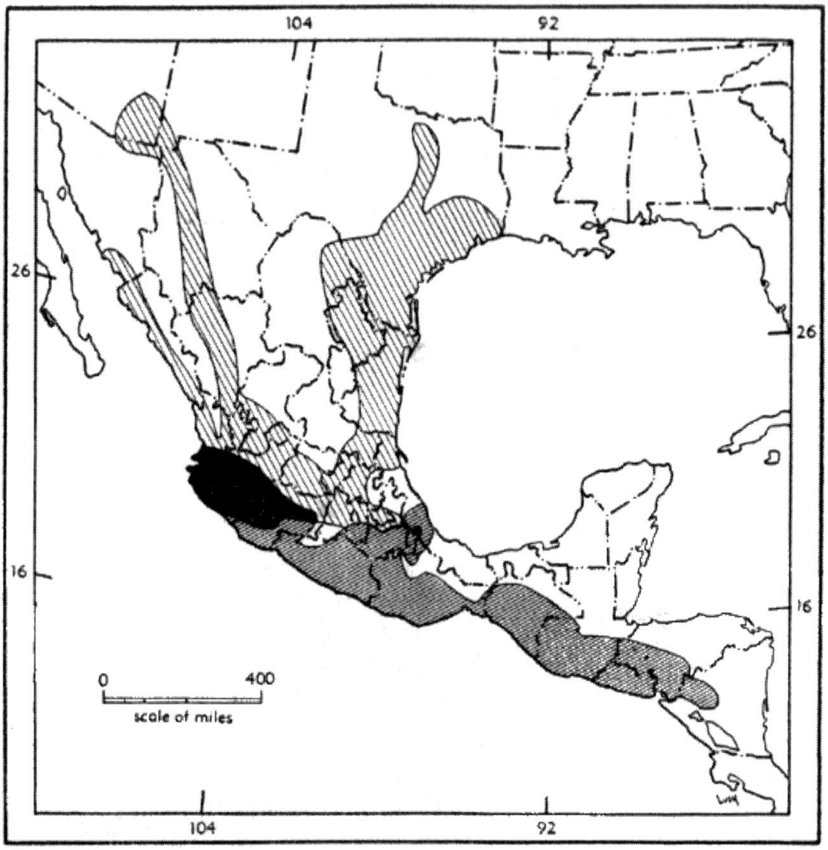

FIG. 9. Geographic distribution of the genus *Baiomys*. Black area shows where the two species occur together. Black dot (Acultzingo, Veracruz) shows locality where *Baiomys taylori* occurs within the range of *B. musculus*, but *B. musculus* is not known to occur at that locality.

SYSTEMATIC ACCOUNT OF SPECIES AND SUBSPECIES

Baiomys musculus

Southern Pygmy Mouse

(Synonymy under subspecies)

Type.--*Sitomys musculus* Merriam, Proc. Biol. Soc. Washington, 7:170, September 29, 1892.

Range.--Southern Nayarit, Michoacan, Mexico, Morelos, Puebla, and central Veracruz, southeastward to western Nicaragua, but unknown from southern Veracruz, Tabasco, and the Yucatan Peninsula (see Figure 10); occurs principally in the arid upper and lower divisions of the Tropical Life-zone.

Characters for ready recognition.--Unless otherwise noted, characters are usable only for the two age-categories of adult and old adult. Differs from *B. taylori* in: hind foot 16 millimeters or more; occipitonasal length, 19 millimeters [Pg 609] or more; zygomatic breadth, 10 millimeters or more; rostrum not deflected ventrally at frontoparietal suture but, instead, curving gradually toward anteriormost point of nasals; cingular ridges and secondary cusps on teeth more pronounced; basihyal having anterior pointed entoglossal process, shoulders of basihyal protruding anteriorly (characteristic of all age categories); baculum having broader shaft, spatulate to knob-shaped tip, wings at base projecting anteriorly; baculum more than 3 millimeters long; short process of incus knob-shaped rather than attenuate; muscular process of posterior crus of stapes prominent.

Characters of the species.--Size large (extremes in external measurements of adults; total length, 100-135; length of tail vertebrae, 33-56; length of hind foot, 14.1-17; length of ear, 9-12); upper parts dark reddish brown, or ochraceous-buff to nearly black; underparts pale pinkish buff to white or pale buffy.

Geographic variation.--Eight subspecies are here recognized (see Figure 10). Features that vary geographically are external size, color of pelage, certain cranial dimensions (occipitonasal length, zygomatic breadth, least interorbital breadth, length of rostrum, length of incisive foramina, depth and breadth of cranium, and alveolar length of upper molar tooth-row).

External and cranial size (except for *B. m. handleyi*) is less in the southernmost subspecies, *B. m. pullus*, *B. m. grisescens*, *B. m. nigrescens*, and more in the northernmost subspecies, *B. m. musculus*, *B. m. brunneus*, and *B. m. infernatis*. Increase in size from south to north is in keeping with Bergman's Rule that within a species, smaller individuals occur in warmer parts of its geographic range. Southern pygmy mice at high altitudes average larger than those from low elevations, except where the two species are sympatric. There the Southern Pygmy Mouse is uniformly larger, regardless of altitude.

Osgood (1909:257, 259) suggested that degree of relative humidity might in some way control color of pelage in both *B. taylori* and *B.*

musculus. In *B. musculus*, the darker subspecies, *B. m. brunneus*, *B. m. nigrescens*, and *B. m. pullus*, occur in zones of rather constant high relative humidity, whereas the paler subspecies *infernatis*, *musculus*, *handleyi*, and to a less extent *grisescens* and *pallidus*, occur in zones of lower relative humidity. This is in keeping with Gloger's Rule, which states that melanins increase in the warm and humid parts of the range of a species, and reddish or yellowish-brown phaeomelanins prevail in arid climates. *B. m. musculus* ranges into areas where relative humidity is such that darker pelages might be expected, but this is in the area where the two species are sympatric, and color of pelage may be an important character of recognition.

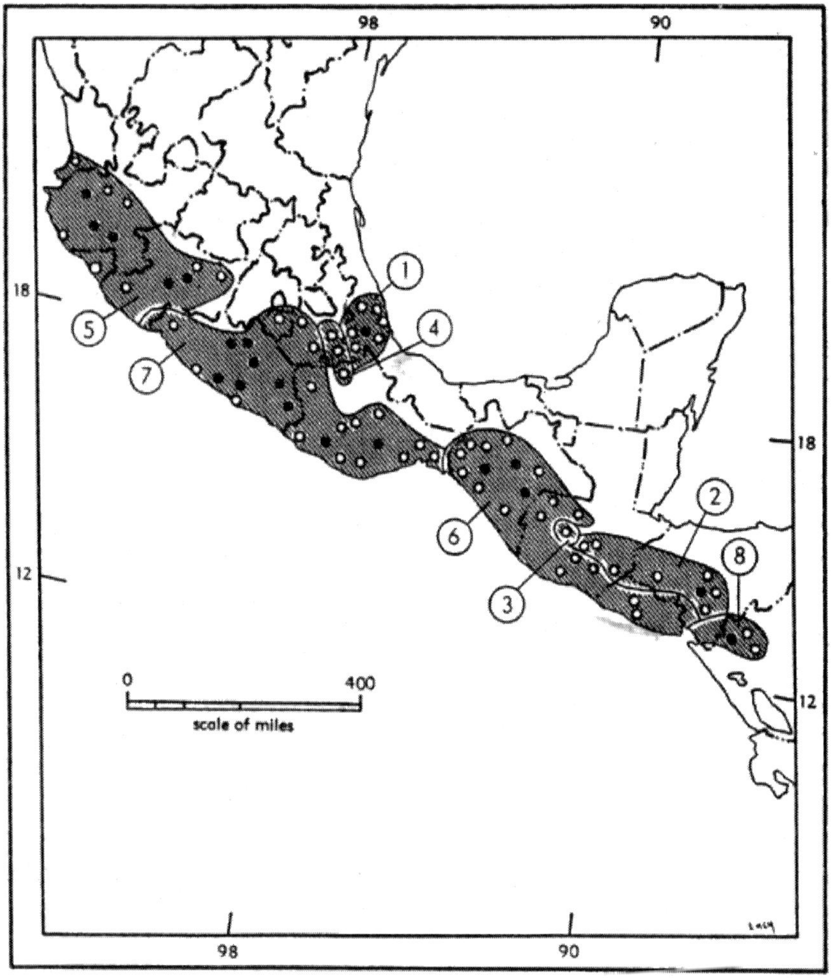

FIG. 10. Distribution of *Baiomys musculus*. Known localities of occurrence are represented by circles and black dots; the former denote localities that are peripheral (marginal) for the subspecies concerned.

1. *B. m. brunneus*
2. *B. m. grisescens*
3. *B. m. handleyi*
4. *B. m. infernatis*

5. *B. m. musculus*
6. *B. m. nigrescens*
7. *B. m. pallidus*
8. *B. m. pullus*

Natural History

Habitat and numbers.--In Veracruz, Dalquest obtained the southern pygmy mouse in stands of tall grass (*Spartina?*) in sandy loam soil bordering, and in, dense vegetation; Davis (1944:394) found the species living in dense stands of grasses and seemingly utilizing underground burrows. Near Chilpancingo, Guerrero, rocky situations seemed to be the preferred habitat. Davis (*loc. cit.*) believed that the species has a wide tolerance to kinds of habitats. In Morelos, Davis and Russell (1954:75) found these mice to be abundant along rock fences separating cultivated fields, and in arid lowlands. In Colima, Hooper (1955b:13) obtained specimens from an open thorn forest in sparse grass and rocky hillside bounding a stream and in litter below shrubs on the floor of a nut-palm forest; in Michoacan, these mice were taken in cane grass, shrubs, and mesquite near an irrigation ditch. From Guatemala, Goodwin (1934:39, 40) records specimens from Sacapulas, a hot, dry, sandy area where cactus and sparse grasses are present, and from La Primavera, on the edges of pine-oak-alder forests. Felten (1958:137) has taken *musculus* from bushy areas in El Salvadore. In 1955, I obtained the southern pygmy mouse 6 mi. SW Izucar de Matemores, Puebla, along a stream in heavy grass bordered by cypress, willow, fig, bamboo, and in rocky grazed area near sugar cane fields.

The southern pygmy mouse seems to be locally abundant in certain parts of its geographic range, and in other parts, scarce. For example, Dalquest (*in. litt.*) recorded the pygmy mouse as common at a place 2 km. N Paraje Nuevo, 1700 feet, Veracruz, where, by means of 50 traps, he took 14 of these mice in one night. The species was scarcer, although the habitat seemed suitable, 3 km. N Presidio, 1500 feet, Veracruz, where he caught only two pygmy mice in several days of trapping. Six miles southwest of Izucar de Matemores, the pygmy mouse was the most common rodent. I have trapped for it in Oaxaca and Veracruz in habitats that seemed almost identical to those mentioned by Dalquest, and also that at Izucar de Matemores, Puebla, with almost no success. The reason for the seeming disparity in numbers at different localities having nearly the same kind of habitat is unknown to me and bears further investigation.

Behavior.--Little is recorded concerning the behavior of this species. David and Russell (*op. cit.*:76) found that of small mammals *B. musculus* was the first to appear at night. I caught mice of this species by hand in the afternoon in Puebla. They seemed to be active from noon until dark. Albert Alcorn wrote in his field notes that specimens were taken near noon at a place 9 mi. NNW Esteli, Nicaragua. My impression is that *musculus* is diurnal to crepuscular.

Enemies and food.--Owl pellets (thought to be those of a barn owl, *Tyto alba*) from within the geographic range of *B. musculus,* from 6 mi. SW Izucar de Matemores, yielded mandibular tooth-rows belonging to *musculus.* Presumably, most of the carnivorous mammals and raptorial birds within the range of the southern pygmy mouse could be listed as enemies. Diurnal to crepuscular habits of this mouse may protect it from some of the nocturnal carnivorous mammals and raptorial birds.

Food of the southern pygmy mouse includes nuts, bark, grass seeds, and leaves. Dalquest (MS) writes that bits of banana proved to be useful bait in trapping these mice in Veracruz.

Reproduction.--Notations concerning lactation and embryos on specimen labels of females suggest that the southern pygmy mouse breeds in all months. I have records of pregnant or lactating females in every month, save January, April, May, and June. The average of 26 counts of embryos or young per litter is 2.92 (1-4).

Baiomys musculus brunneus (J. A. Allen and Chapman)

Peromyscus musculus brunneus J. A. Allen and Chapman, Bull. Amer. Mus. Nat. Hist., 9:203, June 16, 1897; Elliott, Field Columb. Mus. Publ., 105(4):136, July 1, 1905; Elliott, Field Columb. Mus. Publ., 115(8):203, 1907; Osgood, N. Amer. Fauna, 28:259, April 17, 1909.

Baiomys musculus brunneus, Miller, Bull. U. S. Nat. Mus., 79:137, December 31, 1912; Miller, Bull. U. S. Nat. Mus., 128:318, April 29, 1924; Ellerman, The Families and Genera of Living Rodents, 2:402, March 21, 1941; Goldman, Smith. Miscl. Coll., 115:437, July 31, 1951; Goodwin, Bull. Amer. Mus. Nat. Hist., 102:318, August 31, 1953; Miller and Kellogg, Bull. U. S. Nat. Mus., 205:512, March 3, 1955; Booth, Walla Walla Publs., Dept. Biol. Sci., 20:15, July 10, 1957; Hall and Kelson, The Mammals of North America, 2:661, March 31, 1959 (part).

[*Peromyscus musculus*] *brunneus*, Elliott, Field Columb. Mus. Publ., 95(4): 176, 1904.

Peromyscus musculus [*musculus*], Osgood, N. Amer. Fauna, 28:258, April 17, 1909 (part).

Baiomys musculus musculus, Davis, Jour. Mamm., 25:394, December 12, 1944 (part); Goldman, Smith Miscl. Coll., 115:437, July 31, 1951; Hooper, Jour. Mamm., 33:97, February 18, 1952 (part); Hall and Kelson, The Mammals of North America, 2:661, March 31, 1959 (part).

B. [*aiomys*] *m.* [*usculus*] *brunneus*, Hooper, Jour. Mamm., 33:96, February 18, 1952.

Baiomys taylori, Hooper, Jour. Mamm., 33:97, February 18, 1952 (part).

Type.--Adult female, skin and skull; No. 12535/10845 American Museum of Natural History; Jalapa, Veracruz, Republic of Mexico; obtained on April 13, 1897, by F. M. Chapman, original number 1203.

Range.--Central Veracruz, coastal plains and eastern slopes of the plateau of Central Mexico, see Figure 10. Zonal range: Upper Tropical Life-zone (Lowery and Dalquest, 1951:537), parts of the Veracruz and eastern Transverse Volcanic biotic provinces of Goldman and Moore (1945:349). Occurs from near sea level at Boca del Rio, Veracruz, up to 5500 feet 3 km. SE Orizaba.

Diagnosis.--Size medium to large for the species; ground color of dorsum of paratypes near Olive Brown; darkest of specimens of this subspecies examined (from Potrero Viejo, Veracruz) between Prouts Brown and Mummy [Pg 613] Brown; distal two-thirds of guard hairs of dorsum black, proximal third dark gray to sooty; hairs of dorsum black-tipped having subterminal band of Ochraceous-Tawny; sides paler (less of dark brown) than dorsum; venter Deep Olive Buff to clay color, individual hairs pale olive buff at tips, dark gray basally; region of throat and chin sooty gray; ventralmost vibrissae white to base, other vibrissae black to base; ears dark brown, sparsely haired; forefeet and hind feet flesh-colored in palest specimens, sooty in darkest; tail pale brown, slightly paler below than above; presphenoid only slightly constricted towards midline; average and extreme external and cranial measurements of 10 adults from Cerro Gordo, Veracruz, are as follows: total length, 118.9 (112-127); length of tail vertebrae, 45.1 (42-50); length of body, 74.0 (69-78); length of hind foot, 16.0 (16); length of ear from notch, 12.8 (12-13); occipitonasal length, 19.5 (19.0-20.0); zygomatic breadth, 10.3 (10.0-10.8); postpalatal length, 7.1 (6.7-7.5); least interorbital breadth, 3.9 (3.7-4.0); length of incisive foramina, 4.4 (4.1-4.6); length of rostrum, 6.9 (6.5-7.2); breadth of braincase, 9.5 (9.2-9.7); depth of cranium, 7.1 (7.1-7.4); alveolar length of maxillary tooth-row, 3.3 (3.2-3.3); for photographs of skull, see Plate 1*a*, and Plate 3*a*.

Comparisons.--For comparisons with *B. m. nigrescens*, see account of that subspecies. From *B. m. pallidus*, *B. m. brunneus* differs in: dorsal, lateral, and facial coloration deeper reddish brown, more melanins present; venter darker; buff gray rather than whitish buff to gray as in paratypical series; vibrissae black rather than brownish to white; tail sooty, less flesh-colored; forefeet and hind feet averaging slightly grayer; most external and cranial dimensions averaging slightly larger; nasals less attenuated; presphenoid less hour-glass shaped, sides more nearly straight.

From *B. m. infernatis*, *B. m. brunneus* differs in: side of face and neck deep reddish-brown rather than yellowish-gray (the differences in dorsal colorations are greater between *brunneus* and *infernatis* than between *brunneus* and *pallidus*); venter darker buff-gray; tail brownish rather than flesh-colored; forefeet and hind feet average slightly grayer; most external dimensions averaging slightly larger; cranial dimensions nearly the same except length of incisive foramina, which is smaller; presphenoid differs in much the same way as from pallidus.

Remarks.--Specimens from Chichicaxtle, Puente Nacional, 3 km. W Boca del Rio, 1 km. E. Mecayucan, and Rio Blanco (20 km. WNW Piedras Negras), are all paler than the paratypical series and other specimens from within the assigned range of *B. m. brunneus*. All these specimens from the coastal plain average considerably paler than those from the front range and slopes of the mountains. Specimens from Puente Nacional are intermediate in color between paler, grayish brown, specimens from the coastal plains and the darker, brown, specimens from the mountains. When Allen and Chapman (1897:203) described *brunneus*, they did so on the basis of the darker brown mice from the higher altitudes. The name, *brunneus, sensu stricto*, could be restricted to those mice from the higher altitudes of central Veracruz. However, when the mice of intermediate color from Puente Nacional are considered, it seems best to include the material from the coastal plain with *brunneus*. Crania from the higher altitudes are slightly larger than, but not significantly different from, crania of specimens from the coastal plains. Specimens examined from the coastal plains resemble the darker series of *B. m. pallidus* to the west in central Mexico. But there is no evidence of gene flow between the paler coastal specimens and *B. m. pallidus* to the west. In fact, these paler brown mice on the coastal plain grade in color into the darker brown mice from the mountains. The paler mice from the coast may be an incipient subspecies.

The type and paratypes seem to have faded somewhat since they were described by Allen and Chapman (*loc. cit.*) and by Osgood (1909:259). However, the color of the paratypes and other specimens herein assigned is the feature most useful for distinguishing *brunneus* from all other subspecies of *B. musculus*.

Specimens examined.--Total 187 all from VERACRUZ, Republic of Mexico, and distributed as follows: type locality, 4400 ft., 16[1] (including the type), 6[2], 1[3]; *Cerro Gordo*, 1500 ft., 19; *Teocelo* [= *Texolo*], 4500 ft., 1; *2 mi. NW Plan del Rio*, 1000 ft., 14[4]; *Plan del Rio*, 1000 ft., 2[5]; *Carrizal*, 4[2]; Chichicaxtle, 3[2]; *Puente Nacional*, 500 ft., 1[5], 2; *Santa Maria, near Mirador*, 1800 ft., 10[2]; Boca del Rio, 10 ft., 1[5], 8; *Cordoba* [= *Cordova*], 14[1]; *4 km. WNW Fortin*, 4; *Rio Atoyac, 8 km. NW Potrero*, 1; *2 km. N. Paraje Nuevo*, 1700 ft., 9; *El Xuchil, 1 mi. W. Paraje Nuevo*, 6[6]; Potrero Viejo, 1700 ft. 15; *Cautlapan* [= *Ixtaczequitlan*], 4000 ft., 16;

Micayucan, 1; 3 km. SE Orizaba, 5500 ft., 3; Rio Blanco, 20 km. WNW Piedras Negras, 400 ft, 7; *29 km. SE Cordoba, Presidio*, 15[4]; *3 km. N Presidio*, 1500 ft., 2; Presidio, 600 meters, 6[3].

Marginal records.--VERACRUZ: type locality; Chichicaxtle; Boca del Rio, 10 ft.; Rio Blanco, 20 km. WNW Piedras Negras, 400 ft; Presidio; 3 km. SE Orizaba, 5500 ft.

[1] American Museum of Natural History.

[2] U. S. Nat. Museum (Biol. Surv. Coll.).

[3] Chicago Natural History Museum.

[4] Univ. Michigan, Museum of Zoology.

[5] Texas A & M, Coop. Wildlife Res. Coll.

[6] Univ. Illinois, Mus. Nat. History.

Baiomys musculus grisescens Goldman

Baiomys musculus griesescens Goldman, Proc. Biol. Soc. Washington, 45:121, July 30, 1932; Ellerman, The Families and Genera of Living Rodents, 2:402, March 21, 1941; Poole and Schantz, Bull. U. S. Nat. Mus., 178:259, March 6, 1942; Goodwin, Bull. Amer. Mus. Nat. Hist., 79(2):160-161, May 29, 1942 (part); Miller and Kellogg, Bull. U. S. Nat. Mus., 205:513, March 3, 1955 (part); Felten, Senck. Biol., 39:136, August 30, 1958; Packard, Univ. Kansas Publs., Mus. Nat. Hist., 9:401, December 19, 1958; Hall and Kelson, The Mammals of North America, 2:661, March 31, 1959 (part).

Type.--Adult female, skin and skull; No. 257083 U. S. Nat. Mus. (Biol. Surv. Coll.); Comayabuela [= Comayaguela] just south of Tegucigalpa, 3100 feet, Honduras; obtained on March 6, 1932, by C. F. Underwood, original number 838.

[Pg 615]

Range.--Central to south-central Guatemala, east to south-central Honduras. Zonal range: Lower parts of the Merendon Biotic Province of Smith (1949:235). Occurs from 3200 feet at a place 1/2 mi. N and 1 mi. W Salama, Guatemala, up to approximately 4500 feet at Monte Redondo, Guatemala.

Diagnosis.--Size medium to small for the species; general ground color of dorsum between Olive Brown and Buffy Brown; distal fourth of individual guard hairs of dorsum black-tipped, proximal three-fourths gray, underfur black-tipped with subterminal band of Vinaceous-Buff, gray basally; facial region below eye Olive-Buff to Deep Olive-Buff; regions of flanks without black-tipped guard hairs, therefore, appearing paler brownish-buff than dorsum; venter Pale Olive-Buff to whitish in midline, hairs there white to base, laterally grayish basally; hairs in region of throat and chin resemble those of under parts; forefeet and hind feet flesh-colored with grayish suffusion; ears dusky brown; tail almost unicolored, slightly darker brown above than below; coronoid process less acutely falcate than in other subspecies; zygoma bowed. Average and extreme external and cranial measurements of 14 adults from La Piedra de Jesus Sabana Grande, Honduras, are as follows: Total length, 110.7 (100-123); length of tail vertebrae, 44.0 (32-55); length of body, 66.7 (60-70); length of hind foot, 14.1 (12-15); length of ear from notch, 11.8 (10-13); occipitonasal length, 19.3 (18.9-19.8); zygomatic breadth, 10.1 (9.8-10.4); postpalatal length, 6.8 (6.2-7.3); least interorbital breadth, 3.9 (3.8-4.1); length of incisive foramina, 4.3 (4.0-4.5); length of rostrum, 6.9 (6.6-7.2); breadth of braincase, 9.6 (9.2-10.1); depth of cranium, 7.0 (6.8-7.3); alveolar length of maxillary tooth-row, 3.2 (3.0-3.4); for photographs of skull, see Plate 1*b*, and Plate 3*b*.

Comparisons.--For comparisons with *B. m. pullus* and *B. m. handleyi*, see accounts of those subspecies. From *B. m. nigrescens*, *B. m. grisescens* differs in: dorsum less blackish (dark brown to buffy); face buffy below eye rather than brownish-black; venter buffy to whitish in midline, not sooty gray; forefeet and hind feet flesh-colored with gray overtones, not dusky to sooty; zygoma bowed, sides less parallel; braincase and bony palate slightly broader.

Remarks.--Goodwin (1942:160) mentioned that a specimen from the type locality of *grisescens* was as dark as specimens of *B. m. nigrescens* from Guatemala. However, all specimens from Guatemala, other than those from Sacapulas, were referred by Goodwin (1934:40) to *B. m. nigrescens*. My studies reveal a grayish-brown population in central Honduras near to and including the type locality. This population appears to grade into a slightly paler, particularly as concerns color of hind foot and tail, group of Guatemalan mice from 1 mi. S Rabinal, from 1/2 mi. N, 1 mi. E Salama, and from Lake Atescatempa. Specimens from western Guatemala at Nenton and Jacaltenango, on the other hand, are darker brownish-black, more nearly like the paratypical series of *nigrescens* from the Valley of Comitan, Chiapas, Republic of Mexico. This darker brownish-black color of the back persists in specimens from southern Guatemala and El Salvador (see specimens examined of *B. m. nigrescens* for localities), and they are best referred to *nigrescens*. *B. m. grisescens*, in color and certain cranial characters, therefore, seems to grade into two different subspecies: (1) *B. m. handleyi*, pale mice in the Rio Negro valley in central Guatemala, and (2) *B. m. nigrescens*, dark mice from southern Guatemala, and parts of El Salvador.

Felten (1958:136) referred all *B. musculus* from El Salvador to *B. m. grisescens*. Although I have not examined the specimens reported on by Felten (*loc. cit.*), I have examined specimens from Lake Atescatempa, Guatemala (which I refer to *grisescens*), not too distant from Cerro Blanco, and Finca Las Canarias, Department of Ahuachapan, and Laguna de Guija, Department of Santa Ana (localities listed by Felten). It would seem that specimens from these localities might indeed be *grisescens*. However, specimens that I examined from 1 mi. S Los Planes, and 1 mi. NW San Salvador were considerably darker than paratypes of *grisescens* and were nearly intermediate in color between *nigrescens* and *pullus*. I refer the specimens from 1 mi. NW San Salvador, and 1 mi. S Los Planes to *nigrescens* rather than to *grisescens*.

There is no positive evidence that *B. m. grisescens* intergrades with *B. m. pullus* to the south in Nicaragua. But, there is a suggestion that intergradation occurs between these subspecies in a series of 76 skins from La Piedra de Jesus Sabana Grande, Honduras, referable to *grisescens*. A total of 16 of 76 skins from this locality (21 per cent) possess the mid-ventral white stripe found in 18 of 20 skins (90 per cent), from the type locality of *pullus* in Nicaragua. Further collection in areas between central Honduras and western Nicaragua may yield specimens of *B. musculus* that are intermediate in characters between *grisescens* and *pullus*.

Specimens examined.--Total 149, distributed as follows: GUATEMALA: 1 mi. S Rabinal, 3450 ft., 14; 1/2 mi. N, 1 mi. E Salama, 3200 ft., 10; Lake Atescatempa, 10[7]. HONDURAS: Cementario, Gracias, 1[8]; Monte Redondo, 1[8]; El Caliche, Cedros, 1[8]; *La Flor Archaga*, 2[8], 1[9]; Hatillo, 1[8]; *type locality*, 7[8], 6[7] (including the type), 3[9]; *El Zapote, Sabana Grande*, 4[8]; La Piedra de Jesus Sabana Grande, 76[8]; *Cerro de las Cuches Sabana Grande*, 5.

Marginal records.--GUATEMALA: 1/2 mi. N, 1 mi. E Salama, 3200 ft. HONDURAS: El Caliche, Cedros; Hatillo; La Piedra de Jesus Sabana Grande; Cementario. GUATEMALA: Lake Atescatempa; 1 mi. S Rabinal, 3450 ft.

[7] United States National Museum (Biol. Surv. Collections).

[8] American Museum of Natural History.

[9] Univ. Michigan, Museum of Zoology.

Baiomys musculus handleyi Packard

Baiomys musculus handleyi Packard, Univ. Kansas Publs., Mus. Nat. Hist., 9:399, December 19, 1958.

Baiomys musculus musculus, Goodwin, Bull. Amer. Mus. Nat. Hist., 68(1):39-40, December 12, 1934 (part); Miller and Kellogg, Bull. U. S. Nat. Mus., 205:512, March 3, 1955 (part).

Baiomys musculus nigrescens, Hall and Kelson, The Mammals of North America, 2:661, March 31, 1959 (part).

Type.--Adult female, skin and skull; No. 275604 U. S. Nat. Mus. (Biol. Surv. Coll.); Sacapulas, El Quiche, Guatemala; obtained on April 24, 1947, by Charles O. Handley, Jr., original number 991.

Range.--Known only from the type locality in the valley of the Rio Negro. Zonal range: Part of the Chimaltenangan Province of Smith (1949:235).

Diagnosis.--Size medium to large for the species; dorsum Wood Brown in some series to Buffy Brown; guard hairs of dorsum black-tipped, color of underhairs Avellaneous; hairs white to base in region of chin, throat, and median venter; in lateral region, hairs Neutral Gray at base; dorsal surfaces of forefeet and hind feet and ankles white; tail white below, brownish above; nasals truncate anteriorly; frontoparietal suture forming an obtuse angle with the suture separating the parietals; alveolar length of upper molar tooth-row and tail long. Average and extreme external and cranial measurements for nine adults from the type locality are as follows: Total length, 121.4 (115-128); length of tail vertebrae, 50.7

(49-54); length of body, 70.8 (66-77); length of hind foot, 15.3 (15-16); occipitonasal length, 19.6 (18.8-20.7); zygomatic breadth, 10.5 (10.2-11.0); postpalatal length, 6.9 (6.4-7.4); least interorbital breadth, 4.0 (3.9-4.0); length of incisive foramina, 4.2 (4.0-4.5); length of rostrum, 7.2 (7.0-7.7); breadth of braincase, 9.8 (9.7-10.2); depth of cranium, 7.1 (6.8-7.2); alveolar length of maxillary tooth-row, 3.5 (3.4-3.6); for photographs of skull, see Plate 1*c*, and Plate 3*c*.

Comparisons.--From *B. m. nigrescens*, *B. m. handleyi* differs as follows: everywhere paler; forefeet and hind feet whitish instead of dusky to sooty; hairs of anterior part of face white instead of brown; tail bicolored instead of unicolored; anterior tips of nasals truncate rather than rounded; frontoparietal suture forming obtuse angle with suture separating parietals instead of forming right angle; tail and upper molar tooth-row longer.

From *B. m. grisescens*, *B. m. handleyi* differs in: slightly paler above and below, primarily as a result of lacking buff-colored hairs; forefeet and hind feet white, not flesh-colored with gray overtones; tail bicolored, not unicolored; anterior tips of nasals truncate rather than flaring; tail and upper molar tooth-row longer.

Remarks.--*B. m. handleyi* seems to be restricted to the valley of the Rio Negro, in the region of Sacapulas, Guatemala. Stuart (1954:7) points out that the Rio Negro drops down into a gorge at a place near Sacapulas and flows northward through a deep canyon for approximately 60 kilometers. The Rio Negro, then, flows onto the lowlands of the Yucatan Peninsula. The habitat is xerophytic in the valley of the Rio Negro near Sacapulas. Stuart (*op. cit.*:10) suggests that this xerophytic habitat may be continuous to a place to the north of Chixoy, Chiapas, where the vegetation then becomes more mesic. The mesic conditions to the north in Tabasco and Yucatan probably have restricted the movement of pygmy mice to the north. No specimens of this mouse are known from the Yucatan Peninsula or from the State of Tabasco, Mexico. *B. m. handleyi* intergrades with *B. m. grisescens* to the south. Specimens from 1 mi. S Rabinal, and those from a second locality 1/2 mi. N and 1 mi. E Salama, Guatemala, are intermediate in color of pelage between *handleyi* and *grisescens*. Stuart (*op. cit.*:5) mentions the continuity of habitat and tributaries from the Salama Basin into the valley of the Rio Negro. Absence of physiographic and biotic barriers in the corridor between these two basins probably allows for some gene flow between *handleyi* and *grisescens*, and results in populations intermediate in color. To the north and northwest of Sacapulas, the Sierra de los Cuchumatanes rises abruptly and separates the known geographic range of *handleyi* from that of *nigrescens* to the north, while to the west the cactus-mesquite habitat of *handleyi* gives way to the oak-pine timber that, so far as known, does not support *Baiomys*. The difference in elevation and flora seems to restrict gene flow between *handleyi* and the more northern *nigrescens*. The only evidence of integration between these two subspecies is provided by one specimen from Chanquejelve, Guatemala. That specimen is intermediate in color between the pale *handleyi* and blackish-brown *nigrescens*.

The subspecies closest, geographically, to *B. m. handleyi* is *B. m. nigrescens*, from which *B. m. handleyi* differs more in color than from any of the other named subspecies, except *B. m. pullus*. There is a close correlation of pallor of mice and the xeric Rio Negro Valley, and the darkness (melanistic color) of mice and the mesic mountains and valleys to the north.

Specimens examined.--Total 49, from GUATEMALA: type locality, including the type: 12 (U. S. Nat. Mus., Biol. Surv. Coll.), 37 (Amer. Mus. Nat. Hist.).

Baiomys musculus infernatis Hooper

Baiomys musculus infernatis Hooper, Jour. Mamm., 33:96, February 18, 1952; Miller and Kellogg, Bull. U. S. Nat. Mus., 205:512, March 3, 1955; Hall and Kelson, The Mammals of North America, 2:661, March 31, 1959.

Baiomys musculus musculus, Hooper, Jour. Mamm., 28:50, February 15, 1947 (part).

Type.--Adult male, skin and skull; No. 91497 Univ. of Michigan, Museum of Zoology; Teotitlan, Oaxaca, Republic of Mexico, obtained on February 24, 1947, by Helmuth O. Wagner, original number 2702.

Range.--Southeastern Puebla, in the basin drained by the Rio Salado and Rio Quiotepec, into northern Oaxaca. Zonal range: Arid Tropical in a part of the Orizaba-Zempoaltepec Faunal District of the Transverse Volcanic Biotic Province of Moore (1945:218). Occurs from 3100 feet in Oaxaca up to 6000 feet in Puebla.

Diagnosis.--Size medium for the species; dorsum Drab, terminal parts of individual guard hairs black, Neutral Gray basally, distal parts of underfur Pinkish Buff, proximally Neutral Gray; sides same color as dorsum; hairs in region of throat and chin white to base; venter whitish to Neutral Gray with tinges of Pinkish Buff; dorsal parts of forefeet and hind feet whitish with flesh-colored undertones, ventral parts whitish to dusky-gray; tail bicolored, grayish-brown above, white below; tip of tail not bicolored, instead grayish-brown throughout; ears pale brown, sparsely haired; incisive foramina long, not constricted posteriorly. Average and extreme external measurements for 9 adults from the type locality are as follows: total length, 113.9 (106-122); length of tail vertebrae, 44.1 (41-48); length of body, 71.0 (65-79); length of hind foot, 14.8 (13-16); length of ear, 11.9 (11-12). Average and extreme cranial measurements of 7 adults from the type locality are as follows: Occipitonasal length, 20.1 (19.7-20.4); zygomatic breadth, 10.4 (10.2-10.6); postpalatal length, 7.3 (7.0-7.7); least interorbital breadth, 4.2 (4.0-4.4); length of incisive foramina, 4.8 (4.4-5.6); length of rostrum, 7.2 (6.6-7.5); breadth of braincase, 9.6 (9.5-9.8); depth of cranium, 7.4 (7.1-7.6); alveolar length of maxillary tooth-row, 3.3 (3.1-3.4); for photographs of skull, see Plate 1*d*, and Plate 3*d*.

Comparisons.--For comparisons with *B. m. nigrescens* and *B. m. brunneus*, see accounts of those subspecies. From *B. m. pallidus*, *B. m. infernatis* differs in: sides, ears, and dorsum paler (less of dark brown); venter whitish gray rather than gray with tinge of buff and brown; forefeet and hind feet paler; tail bicolored, not unicolored; incisive foramina longer and not constricted posteriorly; mastoid process turning dorsally and sickle-shaped at posteriormost point rather than capitate.

Remarks.--*B. m. infernatis* resembles *B. m. handleyi* more than any other subspecies in color of pelage and in external and cranial dimensions. The resemblance in color between *B. m. pallidus*, in certain parts of its range, and *B. m. handleyi* may have resulted from nearly parallel selective forces that gave rise to two subspecies, widely separated geographically. The same relation obtains between *B. m. infernatis* and *B. m. handleyi*. Both inhabit arid river basins. In them, pale soil and low relative humidity are important passive factors of selection that give adaptive value to the pale colors of pelage of both *infernatis* and *handleyi*.

Specimens from 6½ mi. SW Izucar de Matemores, and 1 mi. SSW Tilapa, Puebla, are intergrades between *B. m. infernatis* and *B. m. pallidus*. These specimens are intermediate in color and cranial characters between the aforementioned subspecies but possess more of the pale brown overtones seen in paratypes of *pallidus*, and are best referred to that subspecies.

Specimens examined (All in Univ. Michigan, Mus. Zool.).--Total 18, all from the Republic of Mexico and distributed as follows: PUEBLA, Tepanaco, 6000 ft., 3, Tehuacan, 5400 ft., 3. OAXACA: Type locality, 3100 ft., 12 (including the type).

Marginal records.--See specimens examined.

Baiomys musculus musculus (Merriam)

Sitomys musculus Merriam, Proc. Biol. Soc. Washington, 7:170, September 29, 1892; Lyon and Osgood, Bull. U. S. Nat. Mus., 62:135, January 15, 1909.

Baiomys musculus [= *musculus*], Mearns, Bull. U. S. Nat. Mus., 56:381, April 13, 1907; Hooper, Jour. Mamm., 36:29, May 26, 1955.

Peromyscus musculus [*musculus*], J. A. Allen and Chapman, Bull. Amer. Mus. Nat. Hist., 9:203, June 16, 1897; Elliot, Field Columb. Mus. Publ., 105(4):135, July 1, 1905; Osgood, N. Amer. Fauna, 28:257, April 17, 1909 (part).

[*Peromyscus*] *musculus*, Trouessart, Cat. Mamm., 1:518, 1898.

[*Peromyscus*] *musculus* [*musculus*], Elliot, Field Columb. Mus. Publ., 95(4):175, July 15, 1904.

Baiomys musculus musculus, Miller, Bull. U. S. Nat. Mus., 79:137, December 31, 1912 (part); Miller, Bull. U. S. Nat. Mus., 128:318, April 29, 1924 (part); Ellerman, The Families and Genera of Living

Rodents, 2:402, March 21, 1941; Poole and Schantz, Bull. U. S. Nat. Mus., 178:258, March 6, 1942; Davis, Jour. Mamm., 25:394, December 12, 1944 (part); Hooper, Jour. Mamm., 28:50, February 15, 1947 (part); Hall and Villa-R., Univ. Kansas Publs., Mus. Nat. Hist., 1:460, December 27, 1949 (part); Hall and Villa-R., Anal. del Inst. Biol., 21:196, September 28, 1950 (part); Goldman, Smith. Miscl. Coll., 115:336, July 31, 1951 (part); Miller and Kellogg, Bull. U. S. Nat. Mus., 205:512, March 3, 1955 (part); Hooper, Occas. Papers Mus. Zool. Univ. Michigan, 565:13, March 31, 1955; Hall and Kelson, The Mammals of North America, 2:661, March 31, 1959 (part).

B. [aiomys] m. [usculus] musculus, Hooper, Jour. Mamm., 33:97, February 18, 1952 (part); Packard, Univ. Kansas Publs., Mus. Nat. Hist., 9:400; December 19, 1958.

Baiomys taylori allex, Hall and Kelson, The Mammals of North America, 2:659, March 31, 1959 (part).

Type.--Adult female, skin and skull; No. 33437/45460 U. S. Nat. Mus. (Biol. Surv. Coll.); Colima (City), Colima, Republic of Mexico, obtained on March 9, 1892, by E. W. Nelson, original number 2055.

Range.--Southwestern Nayarit and northwestern Jalisco, south into Colima, thence eastward into Michoacan. Zonal range: part of arid Lower Tropical Subzone of Goldman (1951:330); approximates part of the Nayarit-Guerrero Biotic Province of Goldman and Moore (1945:349). Occurs from near sea level in Colima up to 5800 feet in Jalisco.

Diagnosis.--Size large for the species; dorsum Olive-Brown in darkest series to Buffy Brown with tones of Fawn Color in the palest series; guard [Pg 621] hairs of dorsum black-tipped, gray basally (in some specimens, guard hairs gray-tipped with subterminal black band, and gray base); underfur of dorsum black-tipped with subterminal band of fawn to buff, Neutral Gray basally; face and head paler than back because of greater number of fawn-colored and buff-colored hairs; hairs on throat and chin white to base; venter and flanks Pale Olive-Buff in palest series to Gray (Pale Gull Gray) in darkest series; individual hairs of venter tipped with white to buff, basally Gray (Dark Gull Gray); forefeet and hind feet white to gray with flesh-colored undertones; tail faintly bicolored, individual hairs above black, below white; nasals flared anteriorly; zygoma and zygomatic plate thick. Average and extreme external and cranial measurements for 8 adults from Armeria, Colima, are as follows: total length, 125.5 (115-135); length of tail vertebrae, 47.5 (42-54); length of body, 75.6 (68-81); length of hind foot, 16.5 (16-17); occipitonasal length, 20.3 (19.8-20.7); zygomatic breadth, 10.7 (10.3-11.1); postpalatal length, 7.4 (7.1-7.7); least interorbital breadth, 4.0 (3.9-4.1); length of incisive foramina, 4.3 (4.1-4.5); length of rostrum, 7.3 (6.9-7.6); breadth of braincase, 9.8 (9.4-10.0); depth of cranium, 7.1 (6.7-7.2); alveolar length of maxillary tooth-row, 3.4 (3.3-3.6); for photographs of skull, see Plate 1*e*, and Plate 3*e*.

Comparisons.--For comparisons with *B. m. brunneus*, *B. m. infernatis*, and *B. m. pallidus*, see accounts of those subspecies. From *B. m. nigrescens*, *B. m. musculus* differs in: dorsum paler throughout (less of blackish brown); region of face and ears paler, more buff and fawn-colored hairs rather than blackish-brown to grayish hairs; vibrissae paler; venter paler, less dark gray and less of sooty-colored undertones, tips of hairs whitish to pale Olive-Buff rather than light gray at tips becoming darker basally; forefeet and hind feet paler, whitish to pale buff-color with flesh-colored undertones, not sooty-colored to dark brown; tail paler below; nasals flaring outward, not tapering toward midline at anteriormost point; zygoma more massive; larger in external and cranial dimensions.

Remarks.--Merriam (1892:170) described *Sitomys* [= *Baiomys*] *musculus* on the basis of 23 specimens (from Colima City, Colima; Armeria, Colima; Plantinar, and Zapotlan, Jalisco). According to the original description, *B. musculus* resembled a small house mouse and was smaller than any known species of *Sitomys* except *S. taylori* [= *Baiomys taylori*]. From *taylori*, *musculus* differed in being larger [in size of body], and in having longer ears and tail, and larger hind feet. When Allen and Chapman (1897:203) described *Peromyscus* [= *Baiomys*] *musculus brunneus* from Jalapa, Veracruz, the specimens described by Merriam from Colima and Jalisco became representative of the nominal subspecies *B. m. musculus*. Osgood (1909:258) assigned specimens from Colima, Guerrero, Jalisco, Michoacan, Morelos, Oaxaca, Puebla, Sinaloa, Veracruz, and Zacatecas to the subspecies *musculus*. Subsequently, Russell (1952:21) named the subspecies *pallidus* from the arid lowlands of Morelos; Hooper (1952:96) described the subspecies *infernatis* from northern Oaxaca and southeastern Puebla; and Goodwin (1959:1) described a new subspecies *nebulosus* from the Oaxaca highlands. Each of the subspecies mentioned immediately above was described from within the geographic range assigned to *B. m. musculus* by Osgood (*loc. cit.*). Hall and Kelson (1959:661) mapped the range of *B. m. musculus* so as to include Colima, parts of Jalisco, Michoacan, Guerrero, Oaxaca, and Veracruz. Lukens (1955:159), in a study of the mammals of Guerrero, has shown that the characters attributed to *B. m. pallidus* are not significantly different from those of pygmy mice studied from Guerrero. He (*loc. cit.*) concluded that: (1) if the specimens of pygmy mice from central Guerrero were typical of the subspecies *musculus*, then *pallidus* did not deserve subspecific recognition, or; (2) the name *B. m. musculus* should be restricted to the larger pygmy mice inhabiting the lowlands immediately adjacent to the Pacific Coast and the area to the north. My data (see Figure 12) show pygmy mice from southwestern Nayarit, northwestern and central Jalisco, Colima, and parts of Michoacan to be significantly larger in certain cranial and external measurements than pygmy mice from Guerrero, Oaxaca, Morelos, and parts of Puebla. This finding essentially corroborates Hooper's (1952a:96) findings. It seems advisable, therefore, to restrict the range of *B. musculus musculus* to the large mice inhabiting west-central Mexico and the coastal lowlands of

Colima and Michoacan. The name *pallidus* is applicable to the smaller mice occupying Morelos, southwestern Puebla, Guerrero, Oaxaca, and southwestern Chiapas.

B. m. musculus intergrades with *B. m. pallidus* in eastern Michoacan and central and western Guerrero. Specimens from San Jose Prura and 12 mi. S Tzitzio, Michoacan, though referable to *B. m. musculus* because of slightly larger size of crania are intermediate in size and color between the smaller and slightly darker *pallidus* to the south and east and the larger, slightly paler *musculus* to the northwest.

Specimens examined.--Total 156 all from the Republic of Mexico, and distributed as follows: NAYARIT: 3 mi. NNW Las Varas, 150 ft., 1. JALISCO: 7 mi. W Ameca, 4000 ft., 2[10]; *6 mi. W Ameca,* 4300 ft., 3[10]; *10 mi. S Ameca,* 5800 ft., 1[10]; *13 mi. S, 15 mi. W Guadalajara,* 3; *13 mi. S, 9 1/2 mi. W Guadalajara,* 1; *3 mi. ENE Santa Cruz de las Flores,* 1; 27 mi. S, 12 mi. W Guadalajara, 1; *4 mi. NE Autlan,* 3000 ft., 5[10]; *Sierra de Autlan,* 5000 ft., 2[10]; *2 1/2 mi. NNE Autlan,* 3000 ft., 8; 2 mi. SSE Autlan, 1; *5 mi. S Purificacion,* 2; Chamela Bay, 1[10]; *2 mi. N La Resolana,* 1500 ft., 6[10]; *1 mi. N San Gabriel,* 4000 ft., 32[10]; 2 mi. N Cuidad Guzman, 5000 ft., 1; 3 mi. E Navidad, 4300 ft., 10[10]. COLIMA: *type locality,* 10[11] (including the type); *3 mi. SE Colima (City),* 5[10]; *4 mi. SW Colima City,* 1; Armeria, 200 ft., 8[11]; *Paso del Rio,* 20[10]. [Pg 623] MICHOACAN: 12 mi. S Tzitzio, 6[10]; San Jose Prura, 4[12]; 1 mi. E, 6 mi. S Tacambaro, 4000 ft., 3[13]; La Salada, 3[11]; 1/2 mi. SE Coalcoman, 15[10].

Marginal records.--NAYARIT: 3 mi. NNW Las Varas, 150 ft. JALISCO: 3 mi. E Navidad, 4300 ft.; 27 mi. S, 12 mi. W Guadalajara. MICHOACAN: 12 mi. S Tzitzio; San Jose Prura; 1/2 mi. SE Coalcoman. COLIMA: Armeria, 200 ft. JALISCO: Chamela Bay.

[10] Univ. Michigan, Museum of Zoology.

[11] U. S. Nat. Museum (Biol. Surv. Coll.).

[12] Chicago Natural History Museum.

[13] Univ. California, Mus. Vert. Zoology.

Baiomys musculus nigrescens (Osgood)

Peromyscus musculus nigrescens Osgood, Proc. Biol. Soc. Washington, 17:76, March 21, 1904; Elliot, Field Columb. Mus. Publ., 105(4):136, July 1, 1905; Lyon and Osgood, Bull. U. S. Nat. Mus., 62:135, January 15, 1909; Osgood, N. Amer. Fauna, 28:259, April 17, 1909.

Baiomys musculus nigrescens, Miller, Bull. U. S. Nat. Mus., 79:137, March 31, 1912; Miller, Bull. U. S. Nat. Mus., 128:318, April 29, 1924; Goodwin, Bull. Amer. Mus. Nat. Hist., 68(1):40, December 12, 1934; Ellerman, The Families and Genera of Living Rodents, 2:402, March 21, 1941; Poole and Schantz, Bull. U. S. Nat. Mus., 178:259, March 6, 1942; Hooper, Jour. Mamm., 28:50, February 15, 1947; Goldman, Smith. Miscl. Coll., 115:357, July 31, 1951; Miller and Kellogg, Bull. U. S. Nat. Mus., 205:513, March 3, 1955; Booth, Walla Walla Publs., Dept. Biol. Sci., 20:15, July 10, 1957; Hall and Kelson, The Mammals of North America, 2:661, March 31, 1959 (part).

[*Peromyscus musculus*] *nigrescens*, Elliot, Field Columb. Mus. Publ., 95(4):176, 1904.

B. [*aiomys*] *m.* [*usculus*] *nigrescens*, Goodwin, Bull. Amer. Mus. Nat. Hist., 79(2):160, May 29, 1942; Hooper, Jour. Mamm., 33:97, February 18, 1952 (part); Packard, Univ. Kansas Publs., Mus. Nat. Hist., 9:399, December 19, 1958.

B. [*aiomys*] *m.* [*usculus*] *musculus*, Booth, Walla Walla Publs., Dept. Biol. Sci., 20:15, July 10, 1957 (part).

Type.--Adult female, skin and skull; No. 76827 U. S. Nat. Mus. (Biol. Surv. Coll.); Valley of Comitan, Chiapas, Republic of Mexico, obtained on December 9, 1895, by E. W. Nelson and E. A. Goldman, original number 8719.

Range.--Southern coastal region and eastern parts of Chiapas, southeastward into central and southern Guatemala, thence south into El Salvador (see Figure 10). Zonal range: parts of Lower Austral; also occurs in parts of the arid division of the Upper Tropical Life-zone, and in parts of the arid division of the Lower Tropical Life-zone; approximates a part of the Chiapas Highlands Biotic Province of Goldman and Moore (1945:349), and parts of the Guatemalan Subregion of Smith (1949:235).

Diagnosis.--Size medium to small for the species; dorsum Vandyke Brown mixed with blackish, individual hairs black-tipped with a subterminal band of Warm Buff, Neutral Gray at base; guard hairs of dorsum black distally, Neutral Gray basally; hairs on sides grayish-brown, facial region like dorsum; chin buffy-brown; vibrissae brown, ventrally some white; venter creamy-buff to grayish, individual hairs creamy-buff at tips, gray basally; in region of throat and chin, hairs tipped with Ochraceous-Buff; dorsal surface of forefeet and hind feet dull whitish gray to brownish-black; tail indistinctly bicolored, dusky above, grayish to brownish below; incisive foramina short, wide medially; [Pg 624] average and extreme external and cranial measurements of 15 adults from 6 mi. NW Tonala, Chiapas, are as follows: total length, 107.5 (100-116); length of tail vertebrae, 41.1 (33-48); length of body, 66.1 (62-73); length of hind foot, 15.0 (14-16); length of ear, 10.9 (10-12); occipitonasal length, 18.9 (18.4-19.7); zygomatic breadth, 9.8 (9.4-10.2); postpalatal length, 6.9 (6.6-7.4); least interorbital breadth, 3.7 (3.5-3.8); length of incisive foramina, 4.4 (4.1-4.8); length of rostrum, 6.7 (6.1-7.1); breadth of braincase, 9.2 (9.0-9.4); depth of cranium, 6.9 (6.5-7.3); alveolar length of maxillary tooth-row, 3.1 (2.9-3.2); for photographs of skull, see Plate 1*f*, and Plate 3*f*.

Comparisons.--For comparisons with *B. m. handleyi*, *B. m. grisescens*, *B. m. musculus*, *B. m. pallidus*, and *B. m. pullus*, see accounts of those subspecies.

From *B. m. brunneus*, *B. m. nigrescens* differs in: dorsum blackish-brown rather than reddish to ochraceous brown; face and ears brownish-black rather than brownish with tinges of ochraceous; vibrissae darker; forefeet and hind feet darker; venter with more grayish tones; dorsalmost part of zygomatic plate projects farther anteriorly; interparietal oval to diamond-shaped and narrower anteroposteriorly; zygomata narrower at anteriormost part; slightly smaller in most cranial and external measurements.

From *B. m. infernatis*, *B. m. nigrescens* differs in: dorsum darker; region of face and ears darker; venter buffy to gray rather than whitish-buff; vibrissae darker; forefeet and hind feet darker; tail darker above and below; incisive foramina shorter, more constricted laterally; cranium slightly smaller in most dimensions.

Remarks.--Hooper (1952a:93-94) reported specimens from the coastal strip of southern Chiapas as the most intensely pigmented, whereas, specimens from central and western Chiapas were distinctly paler. Crania of specimens from the coastal region of southern Chiapas were smaller than crania from the central highlands and mountains of

Chiapas. My studies essentially corroborate the findings of Hooper. The gradation of color between the pale brown *pallidus* to the north in Oaxaca, and the brownish-black *nigrescens* to the south in Chiapas is extremely gradual. Specimens from the central and western parts of Chiapas (see Figure 10 for localities) are difficult to assign to either *pallidus* or *nigrescens*. Equal justification exists for assignment to either subspecies. I have assigned the specimens to *nigrescens* because they are geographically closer to the type locality of *nigrescens*. Specimens from Reforma, Oaxaca (assigned by Hooper, 1952a:93-94, to *nigrescens*), are nearly identical in size and color to paratypes of *pallidus*. I assign the Reforma specimens to *pallidus*.

The darkest of all the specimens examined and assigned to *nigrescens* are from 1 mi. NW San Salvador and 1 mi. S Los Planes, El Salvador. The variations in color in this subspecies closely correspond to degree of relative humidity; the palest samples are from areas of low relative humidity and the darkest are from areas of high relative humidity. In view of the present state of differentiation of specimens from the southern coastal areas of Chiapas and mountainous areas of El Salvador, it would seem that populations there might be incipient subspecies.

Specimens examined.--Total 319. CHIAPAS: *17 mi. W Bochil*, 1[14]; *15 mi. W Bochil*, 1[14]; *14 mi. W Bochil*, 1[14]; Bochil, 6[15]; Ocuilapa, 3500 ft., 5[16]; *5 mi. NNW Tuxtla Gutierrez*, 9; *11 km. W Tuxtla Gutierrez*, 800 m., 2[15]; *10 km. W Tuxtla Gutierrez*, 800 m., 2[15]; *Tuxtla Gutierrez*, 2600 ft., 8[16], 11; *Ocozocoautla*, 10[15], 2[16]; 25 mi. E Comitan, Las Margaritas, 1250 m., 5[17], 24[15]; Cintalpa, 555 m., 1[14], 18[15], 3[17]; *Jiquilpilas*, 2000 ft., 1[16]; San Bartolome, 3[16]; *type locality*, 5700 ft., 26[16] (including the type); 15 mi. SW Las Cruces, 1; Villa Flores, 600 m., 12[15]; *23 mi. S Comitan*, 1[14]; *15 mi. S, 2 mi. E La Trinitaria*, 4; *30 mi. S Comitan*, 2[14]; 35 mi. S Comitan, 1[14]; *3 mi. E Arriga*, 1[14]; 6 mi. NW Tonala, 19; *Tonala*, 8[16]; *Los Amates*, 1[14]; Pijijiapan, 10 m., 7[15]; Mapastepec, 45 m., 25[15], 4[17].

GUATEMALA: Chanquejelve, 1[14]; *Nenton*, 3000 ft., 1[16]; Jacaltenango, 5400 ft., 8[16]; La Primavera, 5[14]; 4 mi. S Guatemala City, 4700 ft., 3; *5 mi. S Guatemala City*, 4050 ft., 10; *6 mi. S Guatemala City*, 4680 ft., 1; *Lake Amatitlan*, 4500 ft., 13[16]; El Progresso (Distrito Santa Rosa), 3[15]; *2 mi. N, 1 mi. W Cuilapa*, 2980 ft., 1[14]; *1 mi. WSW El Molino (Distrito Santa Rosa)*, 2; *2 1/2 mi. W, 2 1/4 mi. N San Cristobal*, 2900 ft., 1; El Zapote, 1[15].

EL SALVADOR: 1 mi. NW San Salvador, 29; 1 mi. S Los Planes, 15.

Marginal Records.--CHIAPAS: Bochil; 25 mi. E Comitan, Las Margaritas, 1250 ft. GUATEMALA: Chanquejelve; La Primavera; Jacaltenango, 5400 ft.; 4 mi. S Guatemala City, 4700 ft.; El Progresso. *El Salvador*: 1 mi. NW San Salvador; 1 mi. S Los Planes. GUATEMALA: El Zapote. CHIAPAS: Mapastepec, 45 m.; Pijijiapan, 10 m.; 6 mi. NW Tonala; 15 mi. SW Las Cruces; Cintalpa, 555 m.; Ocuilapa, 3500 ft.

[14] American Museum of Natural History.

[15] Univ. Michigan, Museum of Zoology.

[16] U. S. Nat. Museum (Biol. Surv. Coll.).

[17] University of Florida Collections.

Baiomys musculus pallidus Russell

Baiomys musculus pallidus Russell, Proc. Biol. Soc. Washington, January 29, 1952; Davis and Russell, Jour. Mamm., 35:75, February 10, 1954; Miller and Kellogg, Bull. U. S. Nat. Mus., 205:512; Hall and Kelson, The Mammals of North America, 2:662, March 31, 1959.

Peromyscus musculus brunneus, Elliot, Field Columb. Mus. Publ., 115(8):203, 1907 (part).

Peromyscus musculus [*musculus*], Osgood, N. Amer. Fauna, 28:257, April 17, 1909 (part).

Baiomys musculus musculus, Miller, Bull. U. S. Nat. Mus., 79:137, December 31, 1912 (part); Miller, Bull. U. S. Nat. Mus., 128:318, April 29, 1924 (part); Davis, Jour. Mamm., 25:394, December 12, 1944 (part); Hooper, Jour. Mamm., 28:50, February 15, 1947 (part); Goldman, Smith, Miscl. Coll., 115:336, July 31, 1951 (part); Miller and Kellogg, Bull. U. S. Nat. Mus., 205:512, March 3, 1955 (part); Booth, Walla Walla Publs., Dept. Biol. Sci., 20:15, July 10, 1957 (part); Hall and Kelson, The Mammals of North America, 2:661, March 31, 1959 (part); Goodwin, Amer. Mus. Novitates, 1929:1, March 5, 1959.

B. [*aiomys*] *m.* [*usculus*] *musculus*, Hooper, Jour. Mamm., 33:97, February 18, 1952 (part).

B. [*aiomys*] *m.* [*usculus*] *nigrescens*, Hooper, Jour. Mamm., 33:97, February 18, 1952 (part).

Baiomys musculus nebulosus Goodwin, Amer. Mus. Novitates, 1929, March 5, 1959.

Type.--Adult female, skin and skull; No. 4501 Texas A&M Cooperative Wildlife Collection; 12 kms. NW Axochiapan, 3500 feet, Morelos, Republic of Mexico, obtained on July 28, 1950, by W. B. Davis, original number 5112.

Range.--Guerrero thence eastward into Morelos and west central Puebla along the southern edge of the Transverse Volcanic Biotic Province (Goldman and Moore, 1945:349), south into Oaxaca, see Figure 10. Zonal range: largely Arid Lower Tropical Subzone of Goldman (1951:330). Occurs from near sea level in Oaxaca and Guerrero up to 6550 feet in Oaxaca.

Diagnosis.--Size medium for the species; dorsum Buffy Brown in palest series to Olive-Brown in darkest series, individual hairs Warm Buff, Neutral Gray basally, some with black tips and a subterminal band of Warm Buff, guard hairs of dorsum black-tipped, gray basally; hairs on sides creamy-buff, gray basally; face same color as back fading to white on throat; vibrissae white-tipped, pale brown basally; venter, whitish with tinges of buff on lower throat, individual hairs having tips white to buffy-white, light gray basally; dorsal surface of forefeet and hind feet whitish to flesh-color; tail indistinctly bicolored, brownish above, grayish brown below; zygoma bowed as in *B. m. grisescens*; tail short; average and extreme external and cranial measurements for 17 adults from Tehuantepec, Oaxaca, are: total length, 117.3 (110-126); length of tail vertebrae, 46.9 (41-51); length of body, 70.4 (65-76); length of hind foot, 15.8 (15-16); occipitonasal length, 18.9 (18.2-20.1); zygomatic breadth, 10.1 (9.7-10.6); postpalatal length, 6.9 (6.6-7.5); least interorbital breadth, 3.8 (3.6-3.9); length of incisive foramina, 4.4 (4.2-4.7); length of rostrum, 6.7 (6.3-7.2); breadth of braincase, 9.3 (8.7-9.7); depth of cranium, 6.6 (6.4-6.8); alveolar length of maxillary tooth-row, 3.2 (3.1-3.4); for photographs of skull, see Plate 1*g*, and Plate 3*g*.

Comparisons.--For comparisons with *B. m. brunneus* and *B. m. infernatis*, see accounts of those subspecies.

From *B. m. musculus*, *B. m. pallidus* differs in: dorsum more olive-gray and brown, less ochraceous on either side of mid-dorsal region; face below eye grayish, not buffy; sides gray with buffy overtone, not creamy with light yellow overtones; venter grayish-white rather than an olive-buff; zygomata more tapering anteriorly; maxillary part of zygoma narrower when viewed from above; external and cranial dimensions smaller.

From *B. m. nigrescens*, *B. m. pallidus* differs in: dorsum paler, fewer black hairs medially; face paler, less sooty; vibrissae brownish with white tips rather than black with brownish tips; venter paler; dorsal surface of forefeet and hind feet whitish to flesh-colored rather than sooty to dusky-white; tail paler; nasals slightly more attenuated; averaging slightly larger in external and cranial measurements.

Remarks.--Russell (1952:21) described *pallidus*, on the basis of specimens from the arid Balsas Basin, of Morelos, as pale gray dorsally. After examining the original material from Morelos, I find the dorsal color of *pallidus* to be much closer to a buffy brown than a pale grayish. Even so, smaller size differentiates *pallidus* from *musculus*. *B. m. infernatis*, not *B. m. pallidus*, is the most pallid of all named subspecies of *B. musculus*.

B. m. pallidus intergrades to the northwest with *B. m. musculus*, to the northeast with *B. m. infernatis*, and to the southeast with *B. m. nigrescens*.

According to Goodwin (1959:2), *B. m. nebulosus* (named on the basis of one specimen) differs from *B. m. musculus* [= *pallidus*] from southern Oaxaca in: darker and longer pelage; larger skull; interorbital region broader and less constricted posteriorly. From *B. m. nigrescens* and *B. m. brunneus*, *B. m. nebulosus* differs as follows: pelage longer and softer; skull larger.

Study of specimens of *B. musculus* from Oaxaca reveals considerable variation in external and cranial measurements as well as color, corresponding to that reported by Goodwin (*loc. cit.*). Specimens from higher altitudes average somewhat darker and larger in external and cranial size than those at lower elevations. These differences seem to be microgeographic and not of subspecific rank. Among specimens that I have studied in Oaxaca are several from different localities (KU 63052, an adult male, from 3 mi. W Miahuatlan; KU 68964, an adult male from 3 mi. W Mitla, 6000 ft.; KU 63055, an adult female from 3 mi. S Candelario, 1200 ft.) that, according to Goodwin (*in. litt.*) match *nebulosus* in reported color, size of body and skull (except for the region of the rostrum).

Two of the three specimens (KU 63052 and 63055) are the darkest of a series in which the palest are inseparable from *B. m. pallidus*. Goodwin, who kindly compared the three specimens with the type of *nebulosus*, mentioned (*in. litt.*) that the skull of the type has a slenderer rostrum. Included in the series of skulls of *B. m. pallidus* from 3 mi. W Mitla are several adults (not seen by Goodwin) with slender rostra. *B. m. nebulosus* is judged to be a synonym of *B. m. pallidus*.

Populations of pygmy mice occurring in partially isolated areas of highland in Oaxaca seem to me to be incipient subspecies.

Specimens examined.--Total 824 all from the Republic of Mexico and distributed as follows: PUEBLA: 2 mi. S Atlixco, 5800 ft., 1; *1 mi. SSW Tilapa*, 5800 ft., 2; *6 mi. SW Izucar de Matemores*, 7; *Piaxtla*, 3900 ft., 4[18]; Acatlan, 4100 ft., 1. MORELOS: 5 mi. W Tepoztlan, 6000 ft., 7[19]; *1 mi. W Tepoztlan*, 6000 ft., 9[19]; *2 mi. SW Tepoztlan*, 7000 ft., 1[20]; *Cuernvaca*, 9[19]; *6 mi. W Yautepec*, 4500 ft., 1[20]; *Yautepec*, 12[19]; *3 mi. N Alpuyeca*, 4000 ft., 2[20]; *Puente de Ixtla*, 2[19]; *Tetecala*, 4[21]; *2 km. S Jonacatepec*, 4500 ft., 6[20]; *type locality*, 6 (including the type). GUERRERO: *Yerbabuena*, 1800 m., 1; *Cueva de tia Juana* [= *1.5 km. SSW Yerbabuena*], 1; *Laguna Honda*, 1840 m. [= *1.5 km. S Yerbabuena*], 3; 9 mi. SE Taxco, 3800 ft., 1[22]; *17 km. S Taxco*, 4000 ft., 2[20]; *Iguala*, 5[19]; *3.2 km. SSE Iguala*, 970 m., 1; 1 km. SSE Texcaizintla, 1600 m., 2; *Teloloapan*, 20[19], 5[24]; *1 km. N Chapa*, 1470 m., 6; *Chapa*, 1470 m., 5; El Limon, 3[18]; 2½ mi. W Mexcala, 2100 ft., 1[20]; *Rio Balsas*, 1[18]; Ayusinaha [= Ayotzinapa], 1[18]; *Tlapa*, [Pg 628] 3900 ft, 1[18]; *2.5 mi. S Almolonga*, 5600 ft., 13[20]; *1 km. N Zihuatanejo*, 1; Zihuatanejo Bay, 4[19]; *Las Gatas* [= *2 km. S. Zihuatanejo*], 2; *2 km. SSE Zihuatanejo*, 9; *4 mi. W Chilpancingo*, 5800 ft., 3[20]; *Chilpancingo*, 4800 ft., 14[18], 21[19], 45[21]; *2 mi. N Tixtla*, 4400 ft., 3[20]; *3.2 km. S Chilpancingo*, 4; *Cd. Chamilpa* [= *12 km. ESE Chilpancingo*], 5; *Tlalixtaquilla*, 4200 ft., 2[18]; *15 km. S. Chilpancingo*, 4300 ft., 10[20]; *1 mi. SW Colotlipa*, 2700 ft., 16[20]; *2 mi. SW Colotlipa*, 2700 ft., 1[20]; *Achuitzotla*, 2800 ft., 7[20]; *8 mi. SW Colotlipa*, 1[20]; *5 mi. S Rincon*, 2600 ft., 2[20]; *8 mi. SW Tierra Colorado*, 600 ft., 1[20]; Rio Aguacatillo, *30 km. N Acapulco*, 1000 ft., 3[20]; 5 mi. ESE Tecpan, 50 ft., 9; *Ejido Viejo, 12 km. NNW Acapulco*, 1; *2 mi. NNW Acapulco*, 7; Acapulco, 3[18], 3[21]; Omentepec, 200 ft., 7[18]. OAXACA: *4 mi. E Huajuapam*, 5000 ft., 1; 2 mi. NW Tamazulapan, 6550 ft., 1; Yalalag, 3000 ft., 5[18]; *11 mi. NW Oaxaca* [*City*], 1; *Yaganiza*, 3900 ft., 1[18]; Oaxaca [City], 5000 ft., 15, 7[21], 7[19], 5[24]; *3 mi. ESE Oaxaca* [*City*], 30; *4 mi. ESE Oaxaca* [*City*], 5050 ft., 1; *10 mi. SE Oaxaca* [City], 1[22]; *Cerro Ocotepec*, 1[23]; Tepantepec, 9[23]; *1 mi. E Tlacolula*, 5500 ft., 53[19]; *3 mi. W Mitla*, 11; Jalapa, El Campanario, 1[23]; *2 mi. SE Matalan*, 5950 ft., 14; *Lachiguiri*, 2[23]; *Tres Cruces*, 10[23]; *Agua Blanca*, 11[23]; *San Jose*, 1[23]; Reforma, 30[19], 7[21], 10[23], 6[24] *Totolapa*, 1[18]; *Nejapa, 85 km. WNW Tehuantepec*, 500 m., 12[19], 6[24]; *Chicapa*, 2[18]; *Gueladu* [= *Jalapa*], 6[23]; *Juchitan, Laguna Superior*; Manteca, 8[23], 1[23]; San Bartolo, 3000 ft., 1[18]; *Ejutla*, 1400 m., 21[19]; *El Bambita, Tequisitlan* 4[23]; *Mixtequilla*, 2[23]; *Guiencola*, 5[23]; *Tehuantepec*,

200 ft., 26[18], 11[19]; *Sola de la Vega*, 26[19], 3[24]; Huilotepec, 13[18], 3[23]; *Santa Lucia*, 24[23]; *Cerro de Paste, Tenango*, 7[23]; *Sta. C. Quieri*, 3[23]; *Santa Marie Ecatepec, Zarzamora*, 13[23]; *Rincon Bamba*, 11[23]; *3 mi. W Miahuatlan*, 5300 ft., 1; *Miahuatlan*, 12[19], 1[23], 6[24]; *San Juan Acaltepec*, 5[23]; *Zapotitlan*, 1[23]; *Llano Grande*, 3[18]; Pinotepa, 700 ft., 2[18]; Juquila, 8[18]; *Arroyo, San Juan, north of Cerro Otate*, 1[23]; Cerro Otate, 3[23]; 3 mi. S Candelaria, 1.

Marginal records.--MORELOS: 5 mi, W Tepoztlan, 6000 ft. PUEBLA: 2 mi. S Atlixco, 5800 ft.; Acatlan, 4100 ft. OAXACA: 2 mi. NW Tamazulapan, 6550 ft; Tepantepec; Oaxaca [City], 5000 ft; Yalalag, 3000 ft; Jalapa, El Campanario; Reforma; Huilotepec; 3 mi. S Candelaria; Cerro Otate; Pinotepa, 700 ft. GUERRERO: Acapulco; Zihuatanejo Bay; El Limon; 9 mi. SE Taxco, 3800 ft.

[18] U. S. Nat. Museum (Biol. Surv. Coll.).

[19] Univ. Michigan, Museum of Zoology.

[20] Texas A & M, Cooperative Wildlife Research Collection.

[21] Chicago Natural History Museum.

[22] California Academy of Sciences.

[23] American Museum of Natural History.

[24] University of Florida Collections.

Baiomys musculus pullus Packard

Baiomys musculus pullus Packard, Univ. Kansas Publs., Mus. Nat. Hist., 9:401, December 19, 1958.

Baiomys musculus grisescens, Goodwin, Bull. Amer. Mus. Nat. Hist., 79(2):161, May 29, 1942 (part); Miller and Kellogg, Bull. U. S. Nat. Mus., 205:513, March 3, 1955 (part); Hall and Kelson, The Mammals of North America, 2:661, March 31, 1959 (part).

Type.--Adult female, skin and skull; No. 71605 University of Kansas Museum of Natural History; 8 mi. S Condega, Esteli, Nicaragua, obtained on July 15, 1956, by A. A. Alcorn, original number 4218.

Range.--West-central Nicaragua, from Matagalpa northwest into the valley of the Rio Esteli, east as far as Jinotega, see Figure 10. Zonal range: Upper Tropical Life-zone.

Diagnosis.--Size medium to small for the species; dorsum Fuscous-Black, individual hairs black-tipped with a subterminal band of Ochraceous-Buff, Neutral Gray at base; some hairs on dorsum all black to Neutral Gray at base; hair on sides Neutral Gray tinged with blackish; face blackish, becoming buffy on sides [Pg 629] of head, and white on throat; vibrissae black; tail unicolored Chaetura Black; forefeet and hind feet sooty to dusky-white; mid-ventral region of venter white, hairs white to base; in region of anus and throat, hairs white-tipped, Neutral Gray at base; average and extreme external and cranial measurements of the type and 16 paratypes are as follows: total length, 117.3 (111-121); length of tail vertebrae, 47.2 (44-50); length of body, 70.4 (66-74); length of hind foot, 15.5 (14-17); length of ear from notch, 11.9 (10-13); occipitonasal length, 19.3 (18.9-19.8); zygomatic breadth, 10.2 (9.7-10.6); postpalatal length, 7.0 (6.8-7.3); least interorbital breadth, 3.9 (3.8-4.1); length of incisive foramina, 4.3 (4.0-4.6); length of rostrum, 7.0 (6.5-7.4); breadth of braincase, 9.6 (9.3-10.0); depth of cranium, 7.0 (6.8-7.3); alveolar length of maxillary tooth-row, 3.1 (3.0-3.2); for photographs of skull, see Plate 1*h*, and Plate 3*h*.

Comparisons.--From *B. m. grisescens*, *B. m. pullus* differs in: dorsum and tail darker; sides and lateral parts of venter grayish instead of buffy-brown, thus forming distinct mid-ventral white stripe; average length of body and tail significantly longer, thus total length greater; maxillary tooth-row significantly shorter; slightly larger in other cranial and external dimensions.

From *B. m. nigrescens*, *B. m. pullus* differs in: dorsum slightly darker; face grayish, not sooty; mid-ventral white stripe (absent in most specimens of *nigrescens*) present and becoming grayish laterally; tail darker, less hairy, and averaging significantly longer; smaller in most external and cranial dimensions.

Remarks.--*B. m. pullus* resembles *B. m. nigrescens* in size and color but can readily be distinguished from *nigrescens* by the shorter tail. *B. m. pullus* intergrades with *nigrescens* as shown by specimens, referable to *B. m. nigrescens*, from 1 mi. NW San Salvador and from 1 mi. S Los Planes, El Salvador. In color of the dorsum, specimens from these localities are intermediate between *nigrescens* and *pullus*.

The mid-ventral white stripe characteristic of *pullus* is present in three of 28 adults from El Salvador. Goodwin (1942:160) reported white hairs on the pectoral region of several topotypes of *B. m. grisescens*. The areas of white hairs on the venter of *grisescens* occur in approximately 10 per cent of the specimens examined, whereas in *pullus*, the frequency of occurrence is 90 per cent. The areas of white hairs in *grisescens* are in broad patches on the pectoral region, while in *pullus*, a white stripe passes from the pectoral region to the inguinal region in both males and females. I know of no selective advantage that the presence of this white stripe would confer on the mice.

Specimens examined.--Total 46, all from NICARAGUA, and distributed as follows: Type locality, 32 (including the type); *9 mi. NNW Esteli*, 8; *8 mi. NNW Esteli*, 3; San Rafael Del Norte, 1 (Amer. Mus. Nat. Hist.); *1 mi. NW Jinotega*, 1; Matagalpa, 1 (Amer. Mus. Nat. Hist.).

Marginal records.--NICARAGUA: San Rafael Del Norte; Matagalpa; type locality.

BAIOMYS TAYLORI

NORTHERN PYGMY MOUSE

(SYNONYMY UNDER SUBSPECIES)

Type.--Hesperomys (Vesperimus) taylori Thomas, Ann. Mag. Nat. Hist., Ser. 5, 19:66, January, 1887.

Range.--Southeastern Arizona and southwestern New Mexico, south into Chihuahua and Durango, just east of the Sierra Madre Occidental, thence southeast through Zacatecas, Aquascalientes, Jalisco, Queretaro, and Guanajuato; two fingerlike projections extend northward, one on the west along the coast of Sinaloa into southern Sonora, and the other on the east covering eastern San Luis Potosi, Tamaulipas, eastern Coahuila, Nuevo Leon, into south, southeast, and north-central Texas. Southern margin of range in central Mexico approximates the 19th degree of latitude (see Figure 11). Arid lower and arid upper subdivisions of the Tropical Life-zone in south; principally Lower Sonoran and Lower Austral life-zones in north.

Characters for ready recognition.--Unless otherwise noted, characters are usable for the age-categories of adult and old adult. Differs from *B. musculus* in: hind foot less than 16 millimeters; occipitonasal length less than 19 millimeters; zygomatic breadth less than 10 millimeters; rostrum deflected ventrally at frontoparietal suture rather than curving gradually toward anteriormost point of nasals; cingular ridges and secondary cusps on teeth reduced or absent; basihyal having entoglossal process much reduced or absent, shoulders of basihyal not protruding anteriorly, but more flattened (characteristic of all age categories); baculum having narrower shaft, knob-shaped tip, wings at base projecting laterally, baculum less than 3 millimeters long; short process of incus attenuate; muscular process of posterior crus of stapes reduced.

Characters of the species.--Size small (extremes in external measurements of adults: total length, 87-123; length of tail vertebrae, 34-53; length of hind foot, 12-15; length of ear, 9-12). Upper parts pale drab or reddish-brown to almost black; underparts grayish to cream-buff.

Geographic variation.--Eight subspecies are here recognized (see Figure 11). Features that vary geographically are mostly the same as those that do so in *B. musculus* (see page 609).

External and cranial size is less in *B. t. allex*, the southernmost subspecies, and progressively more in *B. t. paulus*, *B. t. taylori*, *B. t. ater*, *B. t. subater*, *B. t. fuliginatus*, *B. t. canutus*, and *B. t. analogous*. Size is largest in subspecies that occur at higher altitudes. Those subspecies are *B. t. analogous* and *B. t. fuliginatus*. The correlation with Bergman's Rule is less exact in *B. taylori* than in *B. musculus*. It is noteworthy that the smallest subspecies, *B. t. allex*, occurs in the area where the two species are sympatric.

There is close correlation in *B. taylori*, as also in *B. musculus*, of darker pelages with zones of high relative humidity. The subspecies having dark pelages are: *analogous*, *fuliginatus*, and *subater*. The two first-mentioned subspecies occur at high altitudes, and the other, *subater*, occurs in the humid coastal region of Texas. The paler subspecies, *taylori*, *canutus*, and *allex*, occur at lower altitudes. Two subspecies that occur at relatively high altitudes, *ater* and *paulus*, are reddish-brown. The color of pelage in these subspecies resembles the color of soil upon which they live. Blair and Blossom (1948:5) demonstrated close correlation of color of soil with color of pelage in *B. t. ater* by use of an Ives tint photometer.

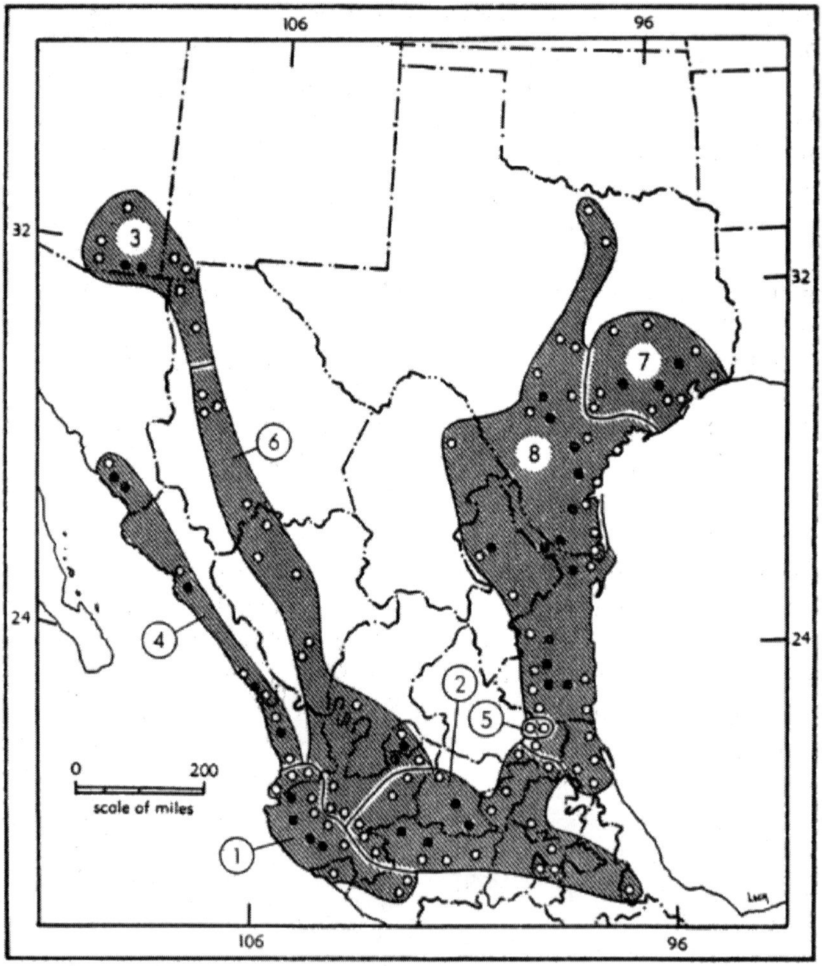

FIG. 11. Distribution of *Baiomys taylori*. Known localities of occurrence are represented by circles and black dots; the former denote localities

that are peripheral (marginal) for the subspecies concerned.

1. *B. t. allex*
2. *B. t. analogous*
3. *B. t. ater*
4. *B. t. canutus*

5. *B. t. fuliginatus*
6. *B. t. paulus*
7. *B. t. subater*
8. *B. t. taylori*

Natural History

Habitat and numbers.--The habitat occupied by the northern pygmy mouse ranges from sparse grassy areas along rock walls in central Mexico (see Davis, 1944:394), and mesquite-cactus associations in southern Texas (Blair, 1952:242) to heavy stands of grasses such as *Bouteloua* sp., *Andropogon* sp., *Hilaria* sp., and sacaton grass intermixed with *Yucca glauca* in New Mexico, Arizona (see Hoffmeister 1956:281), and Chihuahua. Baker (1951:213) reports the species from 2 km. W El Carrizo, Tamaulipas, in dense grass and weeds at the edge of a cornfield. Hooper (1953:7) recorded the northern pygmy mouse in a cultivated field overgrown with herbaceous vegetation at Pano Ayuctle, Tamaulipas. In the State of Sinaloa, Hooper (1955b:13) obtained specimens in grass and among shrubs and vines bordering a fallow field. The northern pygmy mouse, in general, lives in situations more xerophytic and more grassy than does the southern pygmy mouse.

The northern pygmy mouse, as the southern pygmy mouse, is locally abundant in its geographic range. Stickel and Stickel (*op. cit.*: 145) pointed out that on the third night of live-trapping in Bexar County, Texas, there was a sudden increase in unmarked pygmy mice trapped. This increase in numbers, after the resident population was seemingly marked, followed a one-half inch rainfall. Collectors from the University of Kansas, myself included, have had similar experiences in trapping these mice. In the Mexican states of Guanajuato, Queretaro, and Jalisco, *B. taylori* is one of the commonest small mammals. In New Mexico and

Arizona and the Mexican states of Sonora and Sinaloa, nevertheless, these mice are rare.

Stickel and Stickel (*loc. cit.*) thought that the home range normal for *B. taylori* in a grassy habitat was less than 100 square feet, but Blair (1953:10) thought that a complete home range had not been recorded by Stickel and Stickel.

Behavior.--The northern pygmy mouse is crepuscular to nocturnal and where I trapped in northern Mexico was one of the first small rodents to appear in my traps in the evening. Hall and Villa-R (1949:460) recorded this habit in Michoacan. Observations of wild-taken *B. taylori* held in captivity, lend support to its being crepuscular. Captives were rarely active in bright lights, but in diffuse or dim lights the same mice were active.

Blair (1941:381) pointed out that captive *B. t. subater* were much more tolerant of one another than mice of the genus *Peromyscus*. He pointed out also that males aided in care of young. In one litter born in captivity in the course of my study, the female killed the male when the young were four days old. In another instance, the female and two eight-day-old young were killed by the male. Until that time, the male, female, and young had lived together peacefully. In other litters born in captivity, adult males did not harm the other mice.

I have noted, as Blair (*loc. cit.*) did, that *B. taylori* utters high-pitched squeals in a "singing" posture resembling that of the coyote, yet remains silent when being handled.

The northern pygmy mouse makes runways in the grass, in miniature resembling those of *Microtus*, and often uses runways constructed by *Sigmodon*. A small firm nest of finely shredded plant material (mostly grasses) is constructed in burrows or under logs, rocks, or fallen cactus plants. Thomas (1888:447) recorded nests of fine curly grass and cornsilk. Secondary refuge nests are not uncommon. Thomas (*loc. cit.*) states, "If other mice live in the same place, the individuals of *Baiomys* watch till others disappear, then suddenly steal part of the other nest and run to their own with it."

Enemies and food.--Little is recorded of the animals that prey upon the northern pygmy mouse. Twente and Baker (1951:120) found remains of *B. taylori* in 16 per cent of barn owl pellets (*Tyto alba pratincola*) collected 21 mi. SW Guadalajara, Jalisco. Presumably most of the crepuscular and early nocturnal raptorial birds and carnivorous mammals feed on these mice.

Food of *B. taylori* consists in part of grass seeds and leaves, prickly pear (*Opuntia* sp.) and the softer exposed parts of roots of vegetation among which the mice reside.

Reproduction.--The northern pygmy mouse breeds throughout the year. The only months in which I have not recorded pregnant females or females with young are June and October. Forty-one records of embryos or young per litter average 2.48 (less than in *B. musculus*), and range from as few as one to as many as four per litter.

Baiomys taylori allex (Osgood)

Peromyscus allex Osgood, Proc. Biol. Soc. Washington, 17:76-77, March 21, 1904; Elliot, Field Columb. Mus. Publ., 105(6):135, July 1, 1905; Lyon and Osgood, Bull. U. S. Nat. Mus., 62:124, January 15, 1909.

Baiomys taylori allex, Packard, Proc. Biol. Soc. Washington, 71:17, April 11, 1958; Hall and Kelson, The Mammals of North America, 2:659, March 31, 1959 (part).

[*Peromyscus*] *allex*, Elliot, Field Columb. Mus. Publ., 95(4):175, July 15, 1904.

Peromyscus taylori paulus, Osgood, N. Amer. Fauna, 28:255, April 17, 1909 (part).

Baiomys taylori paulus, Miller, Bull. U. S. Nat. Mus., 79:137, December 31, 1912 (part); Miller, Bull. U. S. Nat. Mus., 128:317, April 29, 1924 (part); Ellerman, The Families and Genera of Living Rodents, British Mus. Nat. Hist., 2:402, March 21, 1941 (part); Poole and Schantz, Bull. U. S. Nat. Mus., 178:259, March 6, 1942; Goldman, Smith. Miscl. Coll., 115:373, July 31, 1951 (part); Miller and Kellogg, Bull. U. S. Nat. Mus., 205:512, March 3, 1955 (part); Hall and Kelson, The Mammals of North America, 2:659, March 31, 1959 (part).

Baiomys taylori analogous, Hall and Kelson, Univ. Kansas Publs., Mus. Nat. Hist., 5:367, December 15, 1952 (part).

Type.--Adult male, skin and skull; No. 33429/45452 U. S. Nat. Mus. (Biol. Surv. Coll.); Colima (City), Colima, Republic of Mexico, obtained on March 7, 1892, by E. W. Nelson, original number 2029.

Range.--Colima, western lowlands of Michoacan and Jalisco, thence north into southern half of Nayarit, see Figure 11. Zonal range: arid lower tropical, approximates northern half of the Nayarit-Guerrero Biotic Province of Goldman and Moore (1945:349). Occurs from near sea level in Nayarit, up to 4000 feet in Jalisco.

Diagnosis.--Size small for the species; dorsal ground color pale grayish-brown, near Isabella color; mid-dorsal region washed with blackish, individual guard hairs black to base, other hairs black-tipped with subterminal light olive bands, Neutral Gray at base; laterally, black-tipped hairs less abundant, hairs grayish-white to base; venter Pale Gull Gray to whitish, distal half of individual hairs white, proximal half Neutral Gray; hairs in regions of throat and chin white to base; facial region colored like dorsum, becoming paler below eye; in region of mouth, hairs white to base; dorsalmost vibrissae black to base, others white to base; ears flesh-colored, sparsely haired; tail unicolored, sparsely haired for the species; dark blotches on tail of some series (particularly the paratypical series); dorsal and ventral parts of forefeet and hind feet flesh-colored, whitish to gray in some series. Slightly smaller in most cranial dimensions. Maxillary part of zygoma forming almost a right angle with rostrum, rather than tapering at less than a right angle to rostrum; supraoccipital rounded posteriorly rather than indented on each side of foramen magnum; cranium, relative to length of rostrum, more nearly square; interparietal large relative to size of cranium. Average and extreme measurements of five adults from 2 mi. SSE Autlan are as follows: total length, 100.0 (93-107); length of tail vertebrae, 40.0 (37-44); length of body, 60.0 (56-63); length of hind foot, 14.0 (14); length of ear from notch, 10.5 (10-11); occipitonasal length, 17.3 (16.8-17.9); zygomatic breadth, 9.1 (8.7-9.4); postpalatal length, 6.3 (6.0-6.6); least interorbital breadth, 3.4 (3.3-3.5); length of incisive foramina, 3.9 (3.8-4.0); length of rostrum, 5.5 (5.2-5.8); breadth of braincase, 8.6 (8.0-8.9); depth of cranium, 6.4 (6.0-6.7); alveolar length of maxillary tooth-row, 3.0 (2.8-3.1); for photographs of skull, see Plate 1*i* and Plate 4*a*.

Comparisons.--For comparisons with *B. t. canutus*, see account of that subspecies. From *B. t. analogous*, *B. t. allex* differs in: external and cranial dimensions less; dorsal coloration paler; tail and ears paler and less hairy; dorsum and belly paler; dorsal and ventral parts of forefeet and hind feet paler; median parts of incisive foramina less constricted on either side of midline and wider open laterally; interparietal larger in relation to skull; interorbital breadth greater relative to occipitonasal length.

[Pg 635]

B. t. allex differs from *B. t. paulus* as follows: dorsum gray with yellowish-brown wash rather than fawn to buff; tail unicolored in most series, less hairy; hind feet flesh-colored to light sooty, rather than whitish; rostrum slightly longer relative to occipitonasal length; incisive foramina differ from those of *paulus* in much the same way as from *analogous*.

Remarks.--Osgood (1909:255-256) dismissed as taxonomically unimportant the differences in color of pelage and size of cranium that he observed between the specimens from Colima (City), Colima, representative of *allex* and those representing *paulus* and chose to synonomize *allex* with *paulus*. The differences that Osgood (*loc. cit.*) deemed "... scarcely worthy of recognition ...," are, in fact, not only worthy of recognition, but also important in an understanding of the evolution of *Baiomys taylori* (see speciation p. 659). Recently, I (1958b:17-18) studied ten specimens from Colima (City), Colima, and chose to regard *Peromyscus [= Baiomys] allex* as a subspecies. I suggested (*loc. cit.*) that the geographic range of *B. t. allex* might encompass the southern part of Nayarit, and western Jalisco. Subsequent study of specimens from these areas reveals that the populations there are referable to *allex*.

Most of the specimens obtained from these areas, however, merit special comment.

In color of pelage, those populations from south of the Rio Grande de Santiago and northwest of Guadalajara (4 mi. SE Ahuacatlan; 1 mi. E Ixtlan; Etzatlan) show evidence of intergradation with *paulus* to the east and south (Magdalena, Tequila, and Tala, Jalisco), and with populations more closely adjacent to the south bank of that river. Intergradation is indeed complex in this area. Specimens from some localities seem to be intergrades between *allex* and *paulus*; from other localities, some specimens are referable to *allex*, and the others to *paulus*; from still other localities, all specimens are referable to *allex*.

A series of 39 specimens from 1 mi. SSE Ameca, 4000 ft., Jalisco, are uniformly grayish-brown. This series averages grayer than paratypes of *allex*. There is little, if any, difference between the series from 1 mi. SSE Ameca and paratypes of *allex* in external size of body, hind foot, length of ear, and size and conformation of the cranium. Populations from Ameca and vicinity might be expected to average considerably larger inasmuch as they occur at higher altitudes (see Bergman's Rule, p. 609) then the material from the lower coastal plains to the south in Colima and Michoacan, and at lower elevations in the west in Jalisco and Nayarit. The means of external and cranial measurements are not significantly different between the specimens from the highlands and those from the lowlands. In the area of Ameca where the two species *B. musculus* and *B. taylori* occur together, interspecific competition seems to have limited, perhaps even reduced, size of external and cranial parts of *taylori* (see p. 660).

In color, specimens from the northern part of the valley of the Rio Tepalcatepec (10 mi. S, 1 mi. W Apatzingan) in Michoacan resemble paratypes of *allex*. Intergradation probably occurs to the north with *analogous*.

In the eight specimens from 13 mi. E and 1 mi. N Talpa de Allende, the skull, as reflected in occipitonasal length and zygomatic breadth relative to length of body, is larger than in other specimens here assigned to *allex*. The median part of the belly of the eight specimens is buff-colored rather than whitish-gray as in typical *allex*; the mid-dorsal region also averages darker than in any other specimens referred to *allex*. Additional specimens are needed from this and closely adjacent areas, especially to the west on the coastal plain, in order to determine more accurately the taxonomic status of the mice there. At present, it seems best to refer them to *allex*. Possibly the population represented by the eight specimens is an incipient subspecies.

There is no evidence of hybridization or intergradation of populations of *B. t. allex* with any population of *B. musculus* where the two species occur together.

Specimens examined.--Total 233, all from the Republic of Mexico, distributed as follows: NAYARIT: 3 mi. SE Mirador, 7; 2 *mi. S. Compostela*, 2900 ft., 5; 4 *mi. N Santa Isabel*, 3800 ft., 2[25]; 2 *mi. N Santa Isabel*, 3800 ft., 22[25]; 4 *mi SE Ahuacatlan*, 5200 ft., 2[26]; 1 *mi. E Ixtlan*, 4000 ft., 13[25]; 1 mi. E Ixtlan del Rio, 3700 ft., 1; 2 mi. WNW Valle de Banderas, near sea level, 1. JALISCO: Arroyo de Gavalan, 16[28]; Etzatlan, 6[27]; *Mascota*, 3900 ft., 6[27]; 7 *mi W Ameca*, 15[25]; 6 *mi. W Ameca*, 15[25]; 3 *mi. W Ameca*, 5[25]; Ameca, 4000 ft., 11[27]; 1 *mi. SSE Ameca*, 4000 ft., 38; 2 mi. N Resolana, 1500 ft., 28[25]; 13 mi. E, 1 mi. N Talpa de Allende, 8; 2 mi. SSE Autlan, 5; 1 mi. N San Gabriel, 4000 ft., 1[25]; Las Canoas, 1[28]. COLIMA: Type locality, 10[27] (including the type). MICHOACAN: 9 mi. S Lombardia, 1500 ft., 1; 3 *mi. W Apatzingan*, 1000 ft, 1; Apatzingan, 3[25]; 10 mi. S, 1 mi. W Apatzingan, 800 ft., 10.

Marginal records.--NAYARIT: 3 mi. SE Mirador; 1 mi. E Ixtlan del Rio. JALISCO: Etzatlan; Ameca; 2 mi. N Resolana; Las Canoas. MICHOACAN: 9 mi. S Lombardia; 10 mi. S, 1 mi. W Apatzingan. COLIMA: type locality. NAYARIT: Valle de Banderas.

[25] Univ. Michigan, Museum of Zoology.

[26] California Academy of Sciences.

[27] U. S. Nat. Museum (Biol. Surv. Coll.).

[28] American Museum of Natural History.

Baiomys taylori analogous (Osgood)

Peromyscus taylori analogous Osgood, N. Amer. Fauna, 28:256, April 17, 1909 (part); Elliott, Check-List Mamm., N. Amer. Cont., West Indies and Neighboring Seas, Suppl., Amer. Mus. Nat. Hist., p. 44, January 8, 1917.

Baiomys taylori analogous, Miller, Bull. U. S. Nat. Mus., 79:137, December 31, 1912; Miller, Bull. U. S. Nat. Mus., 128:318, April 29, 1924; Ellerman, The Families and Genera of Living Rodents, British Mus. Nat. Hist., 2:402, March 21, 1941; Poole and Schantz, Bull. U. S. Nat. Mus., 178:259, March 6, 1942; Davis, Jour. Mamm., 25:394, December 12, 1944; Hooper, Jour. Mamm., 28:50, February 15, 1947; Hall and Villa-R., Univ. Kansas Publs., Mus. Nat. Hist., 1:460, December 27, 1949; Hall and Villa-R., Anal. del Inst. Biol., 21:196, September 28, 1950; Goldman, Smith. Miscl. Coll., 114:373, July 31, 1951 (part); Hall and Kelson, Univ. Kansas Publs., Mus. Nat. Hist., 5:367, December 15, 1952 (part); Villa-R., Anal. del Inst. Biol., 23:435, May 20, 1953; Miller and Kellogg, Bull. U. S. Nat. Mus.,

205:512, March 3, 1955; Hooper, Occas. Papers Mus. Zool. Univ. Michigan, 565:13, March 31, 1955; Packard, Proc. Biol. Soc. Washington, 71:17, April 11, 1958.

Peromyscus musculus brunneus, Elliot, Field Columb. Mus. Publ., 115(8):203, 1907 (part).

Peromyscus musculus [*musculus*], Osgood, N. Amer. Fauna, 28:258, April 17, 1909 (part).

Baiomys musculus musculus, Hall and Villa-R., Univ. Kansas Publs., Mus. Nat. Hist., 1:460, December 27, 1949 (part); Hall and Villa-R., Anal. del Inst. Biol., 21:196, September 28, 1950 (part).

Baiomys taylori taylori, Dalquest, Louisiana State Univ. Studies (Biol. Sci. Ser.), 1:155, December 28, 1953 (part); Hall and Kelson, The Mammals of North America, 2:660, March 31, 1959 (part).

Baiomys taylori allex, Hall and Kelson, The Mammals of North America, 2:659, March 31, 1959 (part).

Baiomys musculus musculus, Hall and Kelson, The Mammals of North America, 2:661, March 31, 1959 (part).

Type.--Adult male, skin and skull; No. 120261 U. S. Nat. Mus. (Biol. Surv. Coll.); Zamora, Michoacan, Republic of Mexico, obtained on January 15, 1903, by E. W. Nelson, and E. A. Goldman, original number 15764.

Range.--Central and eastern Jalisco south into Michoacan, east through Guanajuato, Queretaro, thence into Estado Mexico, and Distrito Federal, and west-central Veracruz, see Figure 11. Zonal range: approximately the Transverse Volcanic Biotic Province of Moore (1945:218) and of Goldman and Moore (1945:349). Occurs from 5000 feet, 7 mi. S Ocotlan, Jalisco, up to 8000 feet in Ixtapalapa, Distrito Federal.

Diagnosis.--Size large for the species; dorsum dark Sepia to near blackish medially in freshly taken specimens (Sepia fading to near Fuscous in prepared specimens); belly slaty-gray, hairs Deep Neutral Gray near tips and Dusky Neutral Gray at bases; hairs on back black-tipped with subterminal band of Ochraceous-Tawny (guard hairs blackish to base); hairs of throat and chin white-tipped, gray at bases; dorsal vibrissae black, ventral and anteriormost vibrissae white; hairs on face and sides black-tipped, and Ochraceous-Tawny at base; ears sparsely haired, individual hairs grayish, blackish, and ochraceous; [Pg 638] tail sooty to blackish dorsally, lighter ventrally; forefeet and hind feet sooty brown on dorsal and ventral surface. Skull relatively broad interorbitally; zygoma broad and squared; cranium larger in all dimensions than in most other subspecies. Average and extreme measurements of 10 adults from 1 mi. S, 11 mi. W Zamora, 5400 ft., Michoacan, are: total length, 109.4 (102-121); length of body, 64.3 (58-72); length of tail, 44.9 (39-51); length of hind foot, 14.6 (14-15); occipitonasal length, 18.0 (17.5-18.6); zygomatic breadth, 9.4 (9.1-9.7); postpalatal length, 6.6 (6.2-7.2); least interorbital breadth, 3.5 (3.3-3.8); length of incisive foramina, 4.0 (3.8-4.2); length of rostrum, 6.2 (5.8-6.5); breadth of braincase, 8.7 (8.5-8.9); depth of cranium, 6.6 (6.3-6.9); alveolar length of maxillary tooth-row, 3.1 (3.0-3.3); for photographs of skull, see Plate 2*a* and Plate 4*b*.

Comparisons.--For comparisons with *B. t. allex*, *B. t. canutus*, *B. t. paulus*, and *B. t. fuliginatus*, see accounts of those subspecies. From *B. t. taylori*, *B. t. analogous* differs as follows: sides and dorsum darker, differing most in freshly prepared specimens; dorsal surface of forefeet and hind feet darker; basal part of hairs on belly darker gray; frontal bones less constricted, causing less taper anteriorly in interorbital space; interparietal wider transversely; basioccipital more expanded laterally, narrowing more abruptly at suture between basioccipital and basisphenoid.

Remarks.--The pelage of *analogous* becomes paler with wear as pointed out by Osgood (1909:257). A paratype, U. S. Nat. Mus. 120260, and several specimens from 1 mi. S, 11 mi. W Zamora, Michoacan, are grayish rather than brownish-black. All of these are old adults having the terminal black parts of the hairs on the dorsum nearly worn away. Excluding such grayish individuals, *B. t. analogous*, like *B. t. subater* and *B. t. fuliginatus*, is uniformly brownish-black. Both *analogous* and *fuliginatus* occur in relatively high mountainous country on dark soils or pedregals, and all three of the aforementioned subspecies occur in zones of high relative humidity.

B. t. analogous intergrades with *B. t. paulus* (see account of that subspecies) and *B. t. allex* south and west of Lago de Chapala in Jalisco. Additional specimens are needed from Queretaro and San Luis Potosi in order to ascertain whether or not *B. t. analogous* intergrades with *B. t. fuliginatus* or *B. t. taylori*. Specimens from western Jalisco, in the past referred to *B. t. analogous*, are referable to *B. t. allex* (see account of that subspecies). Specimens obtained west of, and bordering, the Rio del Naranjo in Jalisco show a mixture of characters of both *B. t. allex* and *B. t. analogous*. For example, specimens from 2 mi. N Ciudad Guzman resemble *analogous* on the dorsum, whereas, on the belly, the individual hairs are white-tipped, pale gray at the base, and in over-all appearance are whitish-gray, unlike typical *analogous* (being like *allex* instead). The dorsal surface of the forefeet are sooty to light brownish (as in *analogous*), whereas, the hind feet are flesh-colored (as in *allex*). Another series of specimens from 4 mi. W Leon, Guanajuato, are intergrades between *B. t. analogous* and *B. t. paulus*. These specimens are grayish to brownish on the dorsum, have sooty forefeet and hind feet (more nearly as in *analogous* than in *paulus*), are grayish-white on the venter, and have a distinctly bicolored tail (resembling that of *paulus* more than that of *analogous*). When the average of cranial characters is considered, both series are best referred to *analogous*.

Hooper (1947:50) pointed out that specimens from the pedregal San Geronimo, Distrito Federal, were more nearly black than topotypes and generally showed less brownish hues typical of *analogous*. I have examined this series and several others from this area (see Specimens examined, p. 640) and am convinced that these populations average darker. Actually, the dorsum is more nearly black and the venter is more buffy than in typical *analogous*. The hairs of these individuals average longer than in other populations of *analogous*. Skulls of the specimens from the pedregal are indistinguishable from those of paratypes of *analogous*. The populations from the Distrito Federal seem to be incipient subspecies.

Specimens examined.--Total 696, all from the Republic of Mexico, distributed as follows: SAN LUIS POTOSI: Hacienda Capulin, 5[33]; *3.3 mi. N Tamazunchale, by-road*, 2[34]; 1 mi. N Tamazunchale, 700 ft., 1[35]. VERACRUZ: Acultzingo, 4[29], 1[31]. JALISCO: 1 mi. S Jalostotitlan, 5700 ft., 5; 7 mi. NW Tepatitlan, 3[29]; *6 mi. N, 4 mi. E Tepatitlan*, 6400 ft., 25; *2½ mi. E Tepatitlan*, 6200 ft., 15; *2 mi. S, ½ mi. W Tepatitlan*, 9; *near Tepatitlan*, 2; *5 mi. SW Arrandas*, 6700 ft., 6; *2 mi. E Zapotlanejo*, 23; *2½ mi. E Puente Grande (5½ mi. SW Zapotlanejo)*, 3; *8 mi. S Guadalajara*, 10[29]; *3 mi. ENE Santa Cruz de las Flores*, 9; *4 mi. NE Ocotlan*, 5050 ft., 18; *13 mi. S, 9½ mi. W Guadalajara*, 1; *2 mi. WNW Ocotlan*, 5000 ft., 15; 13 mi. S, 15 mi. W Guadalajara, 2; *Ocotlan*, 5000 ft., 8[30]; *1 mi. S Ocotlan*, 5000 ft., 12; 27 mi. S, 12 mi. W Guadalajara, 9; *1½ mi. N Mazatmitla*, 6[29]; *½ mi. NW Mazatmitla*, 4; *3 mi. WSW Mazatmitla*, 4; 2 mi. N Ciudad Guzman, 5000 ft., 18. GUANAJUATO: 4 mi. N, 5 mi. W Leon, 7000 ft., 25; 5 mi. S Salamanca, 2[29]; *5 mi. E Celaya*, 6000 ft., 6; *1 mi. E Yuriria*, 5725 ft., 3; Salvatierra, 5775 ft., 8; *NE edge Acambaro*, 6050 ft., 10; *Acambaro*, 3[30]. QUERETARO: Toliman, 7[30]; 6 mi. E Queretaro, 6550 ft., 37. HIDALGO: Tula, 2050 m., 1[31]. MICHOACAN: *2 mi. E La Palma, SE side Lago de Chapala*, 7; type locality, 4000 ft., 10[30] (including the type); *9 mi. E Zamora (Camenaro)*, 2[29]; *1 mi. S, 11 mi. W Zamora*, 5400 ft., 17; S Cuitzeo, 36[29]; *Jiquilpan*, 4800 ft., 15; *11 mi. W Jiquilpan*, 6700 ft., 2; *1 mi. E Jiquilpan*, 7; *1 mi. E Zinapecuaro*, 6300 ft., 17; *4½ mi. NE Tarequato (Tarecuato)*, 6600 ft, 1; *Tanganciguaro (Tangancicuaro)*, 5500 ft., 4; *2 mi. N Tarecuato*, 7200 ft., 1; *2 mi. S Maravatio*, 6650 ft, 6; *2 mi. SE Zacapu*, 6600 ft., 11; *1 mi. N Tinquindin (Tinguindin)*, 6300 ft., 2; *3 mi. E Morelia*, 6600 ft., 3; *11 mi. E, 2 mi. S Morelia*, 1; 2 mi. SE Hidalgo (Villa Hidalgo), 6; *1½ mi. N Los Reyes*, 1; *E Los Reyes*, 18[29]; *Los Reyes*, 8[30]; *3 mi. W, 1 mi. N Patzcuaro*, 6600 ft., 2; *N Patzcuaro*, 2[29]; *Patzcuaro* 9[31], 4[30], 4[29]; Uruapan, 1[29]; *E Uruapan*, 12; *2½ mi. E Uruapan (La Presca)*, 2[29]; 2 mi. SW Zitacuaro, 1; 1 mi. E, 6 mi. S Tacambaro, 4000 ft., 11[37]; *La Huacana*, 1[30]. MEXICO: Templo del Sol, Pyramides de San Juan, [Pg 640] Teotihuacan, 8000 ft., 1; *31 km. E Mexico City*, 7500 ft., 11[36]; *17 km. E Mexico City*, 7500 ft, 1[36]; *Cerro La Caldera, 11 mi. ESE Mexico*, 2350 m., 5; 4 km. ENE Tlalmanalco, 2290 m., 9; *Hacienda Cordoba (Cordova)*, 6. MEXICO, D. F.: *Cerro de la Estrella, Ixtapalapa*, 2450 m., 1; *¾ mi. S, 1 mi. E Churubusco*, 2400 m., 2; *5 km. S Mexico City, South of Cd. Universitaria*, 1[32]; *Pedregal San Angel, 2.6 mi. S Monumento a Obregon*, 2; *El Pedregal, 1 km. S San Angel*, 2260 m., 1; *Falda SW Cerro Zacatepec, 3.9 mi. SW Monumento a Obregon*, 1; *2 mi. N Tlalpan, Zacayuca*, 2380 m., 5; *Tlalpan (Pedregal)*, 2400 m., 21[31]; *San Geronimo*, 37[29], 6[38]; *Santa Rosa*, 2700 m., 1[32]; *Tlalpan*, 8; *¾ mi. SW Las Fuentes, Tlalpan*, 2450 m., 25[30]; *Tepepan*, 6[29]; *Rancho La Noria, 1 mi. W Xochimilco*, 2270 m., 4; *500 meters N Xochitepec*, 2250 m., 7; 200 m. N San Mateo Xalpa (Jalpa), 2390 m., 2.

Marginal records.--SAN LUIS POTOSI: Hacienda Capulin; 1 mi. N Tamazunchale. HIDALGO: Tula, 2050 m. MEXICO: Templo del Sol, Pyramides de San Juan, Teotihuacan. VERACRUZ: Acultzingo. MEXICO: 4 km. ENE Tlalmanalco. MEXICO, D. F.: 200 m. N San Mateo Xalpa (Jalpa), 2390 m. MICHOACAN: 2 mi. SW

Zitacuaro; 1 mi. E, 6 mi. S Tacambaro; Uruapan. JALISCO: 2 mi. N Ciudad Guzman; 27 mi. S, 12 mi. W Guadalajara; 13 mi. S, 15 mi. W Guadalajara; 7 mi. NW Tepatitlan; 1 mi. S Jalostotitlan, 5700 ft. GUANAJUATO: 4 mi. N, 5 mi. W Leon. QUERETARO: 6 mi. E Queretaro, 6550 ft.; Toliman.

[29] Univ. Michigan, Museum of Zoology.

[30] U. S. Nat. Museum (Biol. Surv. Coll.).

[31] Chicago Natural History Museum.

[32] American Museum of Natural History.

[33] Museum of Natural History, Louisiana State University.

[34] Univ. Illinois, Mus. Nat. History.

[35] The Museum, Michigan State Univ.

[36] Texas A & M, Cooperative Wildlife Research Collection.

[37] Univ. California, Mus. Vert. Zoology.

[38] University of Florida Collections.

Baiomys taylori ater (Blossom and Burt)

Baiomys taylori ater Blossom and Burt, Occas. Papers Mus. Zool., Univ. Michigan, 465:2, October 8, 1942; Blair and Blossom, Contrib. Lab. Vert. Biol., Univ. Michigan, 40:1, March, 1948; Hoffmeister and Goodpaster, Ill. Biol. Monogr., 24(1):115, December 31, 1954; Miller and Kellogg, Bull. U. S. Nat. Mus., 205:511, March 3, 1955; Hoffmeister, Amer. Midland Nat., 55:281, April, 1956; Packard, Jour. Mamm., 40:146, February 20, 1959; Hall and Kelson, The Mammals of North America, 2:659, March 31, 1959 (part).

Peromyscus taylori paulus, Osgood, N. Amer. Fauna, 28:256, April 17, 1909 (part).

Baiomys taylori [*ater*], Justice, Jour. Mamm., 38:520, November 20, 1957.

Type.--Adult male, skin and skull; No. 85425, University of Michigan, Museum of Zoology; 7 mi. W Hereford, Cochise County, Arizona, obtained on March 25, 1941, by Philip M. Blossom, original number 2195.

Range.--Southeastern Arizona, north to Graham County, thence east to the Animas Valley, Hidalgo County, New Mexico; south to northern Chihuahua and northwest to the southern border of Cochise County, Arizona, see Figure 11. Zonal range: largely lower Sonoran (Apachian Biotic Province of Dice, 1943:56). Occurs from 4300 feet in Chihuahua up to 6200 feet in New Mexico.

Diagnosis.--Size medium for the species; dorsum between Mummy Brown and Prouts Brown; individual tips of hairs intermixture of black and Ochraceous-Tawny, bases of all hairs slate-gray; sides of body and face, Buffy Brown to Cinnamon Brown; belly Cinnamon Buff, proximal half of individual hairs Deep Neutral Gray, distal half white; in region of throat, proximal [Pg 64] fourth of individual hairs gray, distal three-fourths white; dorsal vibrissae black to base, ventral vibrissae white to base; tail brownish above, gray below; dorsal and ventral surface of forefeet and hind feet buffy to gray; interparietal somewhat compressed anteroposteriorly. Average and extreme cranial measurements of 15 adults from 9½ mi. W New Mexico State Line, 5½ mi. N Mexican border, Cochise County, Arizona, are as follows: occipitonasal length, 18.0 (17.5-18.6); zygomatic breadth, 9.5 (9.2-9.9); postpalatal length, 6.6 (6.0-7.1); least interorbital breadth, 3.6 (3.4-3.8); length of incisive foramina, 4.0 (3.8-4.2); length of rostrum, 6.1 (5.7-6.4); breadth of braincase, 8.6 (8.4-9.1); depth of cranium, 6.5 (6.3-6.9); alveolar length of maxillary tooth-row, 3.2 (3.1-3.4). Average and extreme external measurements for six adults from 9 mi. W Hereford, Cochise County, are as follows: total length, 106.3 (98-115); length of tail vertebrae, 42.3 (39-46); length of body, 64 (59-69); length of hind foot, 13.6 (13-14.2); length of ear from notch, 11.1 (10.5-11.5); for photographs of skull, see Plate 2*b*, and Plate 4*c*.

Comparisons.--For comparisons with *B. t. canutus*, see account of that subspecies. From *B. t. paulus*, the subspecies to the southeast, *B. t. ater* differs in: dorsum darker brown; tail less strikingly bicolored; belly buffy rather than whitish to white-gray; forefeet and hind feet darker dorsally and ventrally; posterior margin of basioccipital bowed anteriorly in a broad U-shape with a secondary small median anteriorly directed U-shaped curve, rather than bowed anteriorly in a simple U-shape; interparietal more compressed anteroposteriorly; coronoid process of mandible so acutely recurved that tip of coronoid points posteroventrally and appears sickle-shaped.

Remarks.--Blossom and Burt (1942:1) described *B. t. ater* as the darkest of the known subspecies. It is dark, but specimens from some parts of the ranges of *B. t. analogous*, *B. t. fuliginatus*, and *B. t. subater* exceed in melanins the darkest individuals of *ater*. Blair and Blossom (1948:5) also concluded by the use of an Ives tint photometer that *B. t. subater* was significantly darker than *B. t. ater*.

When paratypes of *ater* and specimens of *B. t. paulus* are compared, the darkest individuals of *ater* exceed but slightly the darkest of *paulus*. The darkest specimens of *paulus* occur in southern Zacatecas, and northern Jalisco, and the palest of the series are in northern Durango and southern Chihuahua. When paratypes of *ater* and *paulus* are compared, the difference in color is readily distinguishable. Specimens from 1½ mi. N San Francisco, in northern Chihuahua, appear to be intermediate in color between *ater* and *paulus* except for a faint tinge of buff ventrally. In characters of the crania, these specimens resemble *ater* and are referred to that subspecies. A slightly different pattern of color is present in pygmy mice from the Peloncillo Mountains and the Animas Valley of New Mexico; the upper parts resemble those of paratypes of *ater*, but the venter has only the faintest suggestion of the buffy wash. Crania of these specimens from New Mexico are inseparable from those of paratypes of *ater*, and the specimens are, therefore, referred to *ater*.

When specimens are arranged by localities from Arizona east into southern New Mexico, thence south into Chihuahua and Durango, gradual intergradation in color is evident from dark in the north to pale browns in the south, whereas, size and shape of interparietal and size and shape of coronoid process of the lower jaw divide quite distinctly into two morphological types in central Chihuahua.

Cranial variation in size and proportion among adults is slight throughout the range of *ater* compared to variation detected in other subspecies of *Baiomys taylori*. Perhaps such a relatively stable pattern of characters of the crania reflects the homogeneity of the gene pool, with respect to these characters, of the populations sampled. The fact that the color of the pelage of this subspecies varies considerable throughout its known range and that the crania do not is perhaps a clue to the mode of inheritance of characters in these mice. Seemingly, color of pelage is inherited independently of characters of the cranium. The relative lack of variability in the crania of *ater* may result from uniform environmental conditions, which have served to select for uniform characters in the populations. All of the other wide-ranging subspecies of *B. taylori* occupy more diverse habitats than *ater*. Secondly, the rather abrupt change in the cline of measured characters of the crania between *ater* and *paulus* in central Chihuahua suggests a secondary zone of intergradation. The probable cessation of gene flow in the past between these two subspecies, allowing *ater* to be isolated for a time, may also, in part, account for the relative lack of variability in the crania of *ater*.

Specimens examined.--Total 58, distributed as follows: ARIZONA: *Graham County*: 1$^1/_2$ mi. SW Ft. Grant, Graham Mts., 1[39]; *Pima County*: 1$^1/_2$ mi. ENE Greaterville, Thurber Ranch, 2[39]; *Santa Cruz County*: Patagonia, 3[39]; *Cochise County*: *9 mi. W Hereford*, 10[43]; type locality, 2[43] (including the type); *5 mi. W Hereford*, 5[43]; 9$^1/_2$ mi. W New Mexico State Line, 5$^1/_2$ mi. N Mexican border, 20[42]; *3 mi. E, 1 mi. N Chiricahua*, 1[42]. NEW MEXICO: *Hidalgo County*: 18 mi. S, 2 mi. W Animas, 2; *22 mi. S, 2 mi. W Rodeo*, 6000 ft., 1[40]; *22 mi. S, 2 mi. E Rodeo*, 6000 ft., 3[40]; 25$^1/_2$ mi. S Animas, 6200 ft. (in Big Bill Canyon), 1[40]. CHIHUAHUA: *5$^1/_2$ mi. N, 2 mi. W San Francisco*, 5100 ft., 1; *2$^1/_2$ mi. N, 3 mi. W San Francisco*, 5200 ft., 1; 1$^1/_2$ mi. N San Francisco, 5100 ft., 4; Casas Grandes, 4300 ft., 1[41].

Marginal records>--ARIZONA: 1$^1/_2$ mi. SW Ft. Grant, Graham Mts. NEW MEXICO: 18 mi. S, 2 mi. W Animas; 25$^1/_2$ mi. S Animas (in Big Bill Canyon). CHIHUAHUA: 1$^1/_2$ mi. N San Francisco; Casas Grandes. ARIZONA: Patagonia; 1$^1/_2$ mi. ENE Greaterville, Thurber Ranch.

[39] University of Illinois, Museum of Natural History.

[40] University of New Mexico.

[41] U. S. Nat. Museum (Biol. Surv. Coll.).

[42] University of Arizona.

[43] Univ. Michigan, Museum of Zoology.

Baiomys taylori canutus, new subspecies

Peromyscus taylori paulus, Osgood, N. Amer. Fauna, 28:255, April 17, 1909 (part).

Peromyscus musculus [*musculus*], Osgood, N. Amer. Fauna, 28:256, April 17, 1909 (part).

Baiomys taylori paulus, Miller, Bull. U. S. Nat. Mus., 79:137, December 31, 1912 (part); Miller, Bull. U. S. Nat. Mus., 128:317, April 29, 1924 (part); Burt, Miscl. Publ., Mus. Zool., Univ. Michigan, 39:54, February 14, 1938; Goldman, Smith. Miscl. Coll., 115:373, July 31, 1951 (part); Miller and Kellogg, Bull. U. S. Nat. Mus., 205:512, March 3, 1955 (part); Hooper, Occas. Papers Mus. Zool., Univ. Michigan, 565:13, March 31, 1955; Hall and Kelson, The Mammals of North America, 2:659, March 31, 1959 (part).

Baiomys musculus musculus, Goldman, Smith. Miscl. Coll., 115:336, July 31, 1951 (part).

Type.--Adult male, skin and skull; No. 62075, University of Kansas, Museum of Natural History; 1 mi. S Pericos, Sinaloa, Republic of Mexico; obtained on June 14, 1954, by A. A. Alcorn, original number 1754.

Range.--Central Nayarit northward through western Sinaloa, to as far north as south-central Sonora, see Figure 11. Zonal range: Lower arid tropical, closely approximating the Sinaloan Biotic Province of Goldman and Moore (1945:349). Occurs from near sea level at Escuinapa (43 feet), Sinaloa, to 3200 feet at a place 2 mi. WNW Tepic, Nayarit.

Diagnosis.--Dorsal ground color Buffy Brown (some specimens near Olive Brown); proximal fourth of individual guard hairs of dorsum black-tipped, distal three-fourths dark grayish; dorsal underfur black-tipped having subterminal band of Buffy Brown; hair around eyes buffy to base; belly Pallid Neutral Gray with overtones of buff; individual hairs in region of chin whitish-gray to bases; vibrissae blackish to bases except ventralmost, those being white to base; tail Dark Olive above, slightly paler below. Average and extreme external measurements of 13 adults from 15 mi. N Rosario, Chele, Sinaloa, 300 ft., are as follows: Total length, 109.6 (99-120); length of tail, 43.4 (38-49); length of body, 66.2 (58-75); length of hind foot, 11.2 (10-12). Average and extreme cranial measurements of 19 adults from the same place are as follows: occipitonasal length, 18.2 (17.7-18.9); zygomatic breadth, 9.6 (9.2-10.1); postpalatal length, 6.9 (6.5-7.3); least interorbital breadth, 3.6 (3.4-3.8); length of incisive foramina, 3.9 (3.5-4.2); length of rostrum, 5.9 (5.5-6.6); breadth of braincase, 8.7 (8.3-8.9); depth of cranium, 6.5 (6.2-6.7); alveolar length of maxillary tooth-row, 3.1 (3.0-3.2); breadth of zygomatic plate, 1.8 (1.6-2.0); for photographs of skull, see Plate 2*c*, and Plate 4*d*.

Comparisons.--From *B. t. ater*, *B. t. canutus* differs in: dorsum slightly grayer; belly whitish to pale-gray with only faint tones of buff, rather than cinnamon-buff to buff-gray; forefeet and hind feet flesh-colored to grayish above instead of whitish to flesh-colored; tail paler above, less hairy, scales more evident; interparietal relatively larger from anteriormost to posteriormost points; incisive foramina tapering less abruptly posteriorly, not constricted towards midline; over-all size of body and cranium somewhat larger.

From *B. t. paulus*, *B. t. canutus* differs in: dorsum grayish-brown rather than fawn-colored (not differing appreciably from extremes of darker brown specimens of *paulus*); forefeet and hind feet flesh-colored to grayish above [Pg 644] rather than white above; tail less hairy, unicolored to faintly bicolored rather than distinctly bicolored; braincase slightly larger; alveolar length of maxillary tooth-row slightly less.

From *B. t. analogous*, *B. t. canutus* differs in: dorsum paler, less of dark brown hues; belly paler; forefeet and hind feet slightly paler, less sooty above; tail less hairy, paler and having scales evident; jugal of zygoma extending ventrally to a point immediately above, instead of below, level of alveolus of upper molars; nasals more nearly truncate anteriorly; infraorbital foramina less deeply notched toward midline of skull; body and skull averaging smaller throughout.

From *B. t. allex*, *B. t. canutus* differs in: dorsal ground color grayish rather than fawn color having grayish overtones; underfur on dorsum darker gray; dorsal surface of forefeet and hind feet flesh-colored to grayish rather than flesh-colored; incisive foramina tapering to a point posteriorly rather than rounded posteriorly; interparietal relatively smaller; body and skull averaging larger throughout.

Remarks.--Burt (1938:54) reluctantly assigned specimens from Ciudad Obregon to *B. t. paulus*, probably being influenced by the resemblance in size. He suggested that, perhaps, a distinct subspecies occurs in the State of Sonora. Study of larger series of specimens than were available to Burt reveals that populations of pygmy mice inhabiting the northwest coastal plains of Mexico are indeed distinct.

The darkest of the material assigned to *canutus* is from Nayarit (for specific localities see specimens examined). According to Tamayo (1949:Carta de Suelos), color of soil changes from chestnut in northern Sinaloa to black in southern Sinaloa and northern Nayarit. There seems, therefore, to be a close correlation between color of pelage and color of soil in this area. In Nayarit, particularly in the central and southern parts, the mice are intermediate in color between the paler, grayer population to the north and the more brownish samples, representative of *allex* to the south. The coastal vegetation changes from the arid tropical thorn forests of the north and central parts of Sinaloa to a savannah in Nayarit, thence to a tropical deciduous forest farther south (see Leopold, 1950:508).

In size and color, specimens from 3 mi. SE Tepic and 2 mi. SW Rosa Morada are intermediate between the larger, grayer *canutus* and the smaller, light-brownish *allex*. In size of cranium, these specimens are more nearly like *canutus*, and are referred to that subspecies. Mice from the western coastal plain are relatively homogeneous as regards size of body and skull, except that those from 13.5 mi. S Acaponeta, Nayarit, average somewhat larger.

B. t. canutus, like *B. t. subater*, is predominantly a lowland or coastal subspecies. The pallor of the former, that lives on generally paler soils, presumably is of adaptive value.

Pygmy mice are seemingly rare in the northern part of the range of this subspecies. J. Raymond Alcorn and Albert Alcorn were successful in collecting only two specimens from the type locality after three successive nights of trapping with 100 traps set each night. Only six specimens are known from Sonora. These were obtained in the irrigated regions of Ciudad, Obregon, and Navajoa. Charles Sibley obtained one specimen 10.6 mi. SE Ciudad Obregon in a "maguey field." I obtained one specimen 1 mi. NNW Navajoa in a sparse grassway, 20 feet wide, bordering an open sewer, which coursed northward into the Rio Mayo. Irrigated wheat fields bordered the grassway and ditch.

Specimens examined.--Total 70 all from the Republic of Mexico and distributed as follows: SONORA: [Ciudad] Obregon, 4[44]; 10.6 mi. SE [Ciudad] Obregon, 1[45]; 1 mi. NNW Navajoa, 1. SINALOA: type locality, 2 (including the type); Culiacan, 175 ft., 2[46]; Mazatlan, 1[48]; *15 mi. N Rosario, Chele*, 300 ft., 35[47]; Rosario, 3[46]; Escuinapa, 5[48]; *Railroad Station Escuinapa*, 43 ft., 2[45]. NAYARIT: Acaponeta, 4[46]; *13.5 mi. S Acaponeta Junction*, 6[49]; 2 mi. SW Rosa Morada, 2; *2 mi. WNW Tepic*, 3200 ft., 1; 3 mi. SE Tepic, 1.

Marginal records.--SONORA [Ciudad] Obregon. SINALOA: type locality; Escuinapa. NAYARIT: Acaponeta; 3 mi. SE Tepic. SINALOA: Mazatlan.

[44] Coll. Univ. California, Los Angeles.

[45] Univ. California, Mus. Vert. Zoology.

[46] U. S. Nat. Museum (Biol. Surv. Coll.).

[47] Univ. Michigan, Museum of Zoology.

[48] American Museum of Natural History.

[49] Univ. Illinois, Mus. Nat. History.

Baiomys taylori fuliginatus, new subspecies

Baiomys taylori taylori, Dalquest, Louisiana State Univ. Studies (Biol. Sci. Ser.) 1:155, December 28, 1953 (part).

Baiomys taylori taylori, Booth, Walla Walla Publs. Dept. Biol. Sci., 20:15, July 10, 1957 (part).

Type.--Adult male, skin and skull; No. 36765, University of Kansas, Museum of Natural History; 10 mi. E, 2 mi. N Ciudad del Maiz, 4000 ft., San Luis Potosi, Republic of Mexico; obtained on January 17, 1950, by J. R. Alcorn, original number 10400.

Range.--Occurs in the Sierra Madre Oriental of the northeastern third of San Luis Potosi. Zonal range: Upper Tropical (see Dalquest, 1953:10); approximates a part of the Sierra Madre Oriental Biotic Province of Goldman and Moore (1945:349, 356). Occurs from 2000 feet at El Salto up to 4000 feet at Ciudad del Maiz.

Diagnosis.--Size large for the species; ground color of dorsum Chaetura Drab; individual guard hairs of dorsum black to base, distal fourth of hairs of underfur in posterior half of dorsum tipped with grayish-brown, proximal three-fourths Dark Neutral Gray; in anterior region of dorsum, posterior to ears, distal third of hairs grayish-brown and proximal two-thirds Dark Neutral Gray to base; sides slightly paler than dorsum; ground color of belly Neutral Gray, individual hairs of belly and throat tipped with Pallid Neutral Gray, basally Deep Neutral Gray to Dark Neutral Gray; tips of individual hairs of face Ochraceous-Tawny; lateral vibrissae whitish, dorsal and ventral vibrissae black to base; forefeet and hind feet sooty above and below, thigh bearing [Pg 646] some white-tipped hairs; tail near Chaetura Drab above, Pale Neutral Gray below; anterior part of jugal projecting slightly ventrally and forming small protuberance at point of articulation with maxillary part of zygoma; jugal extending anteriorly nearly to lacrimal. In most cranial measurements averaging as large as *B. t. analogous*. Average and extreme measurements of the type and three additional paratypes, all adults, are: total length, 105.5 (101-109); length of tail, 39.8 (35-42); length of body, 65.8 (63-68); length of hind foot, 14.3 (14-15); length of ear from notch, 11 (11); occipitonasal length, 18.1 (18.1-18.8); zygomatic breadth, 9.6 (9.3-9.8); postpalatal length, 6.5 (6.0-6.7); least interorbital breadth, 3.4 (3.3-3.6); length of incisive foramina, 4.0 (3.8-4.2); length of rostrum, 6.3 (6.1-6.4); breadth of braincase, 8.8 (8.6-8.9); depth of cranium, 6.7 (6.5-6.8); alveolar length of maxillary tooth-row, 3.2 (3.1-3.3); for photograph of skull, see Plate 2*d*, and Plate 4*e*.

Comparisons.--From *B. t. taylori*, *B. t. fuliginatus* differs in: dorsum slightly darker than in darkest *taylori*; tail densely haired, bicolored rather than unicolored; belly sooty to grayish rather than grayish to whitish; forefeet and hind feet sooty to grayish rather than flesh-colored; incisive foramina less bowed laterally, more nearly straight; interparietal compressed anteroposteriorly, less diamond-shaped.

From *B. t. paulus*, *B. t. fuliginatus* differs in: dorsum dusky to blackish rather than fawn color; belly sooty to grayish rather than buffy to whitish-gray; forefeet and hind feet sooty to grayish rather than whitish; zygoma more nearly forming a right angle with rostrum or skull, less tapered anteriorly; anterior part of jugal possessing ventral projection; jugal extending nearly to lacrimal on posterior surface of maxillary part of zygoma.

From *B. t. analogous*, *B. t. fuliginatus* differs in: mid-dorsal region blacker, less brownish; tail distinctly bicolored rather than unicolored to faintly bicolored; incisive foramina not constricted medially; presphenoid broader (at narrowest point); jugal differs much the same as it does from *paulus*; nasals anteriorly truncate instead of rounded.

Remarks.--Dalquest (1953:155-157) and Booth (1957:15) assigned all of the pygmy mice that they examined from the state of San Luis Potosi to *B. t. taylori*. Examination of all of the material that was available to Dalquest, plus additional specimens at the University of Kansas Museum of Natural History, reveals that there are three subspecies in San Luis Potosi. *B. t. taylori* occurs in the eastern part of the State at lower altitudes; *B. t. analogous* occurs to the southeast at higher altitudes; *B. t. fuliginatus* occurs in the northeastern part of the State in the Sierra Madre Oriental.

Specimens obtained from Ebano, Pujal, and Tamuin, representative of *B. t. taylori*, are much paler on the belly and on the ventral surface of the forefeet and hind feet than are specimens from Ciudad del Maiz, representative of *B. t. fuliginatus*. The tail in *B. t. taylori* is nearly unicolored and less hairy than in the paratypical series of *fuliginatus*.

Specimens from 4 km. NE Ciudad Valles are nearly intermediate in color of the belly, dorsum, forefeet and hind feet, and tail, between the palest mice from the coastal plain and the darker mice in the mountains of the northeastern part of the State (specimens from El Salto average paler, however, than the type and paratypes). These specimens seem to be intergrades between *B. t. taylori* to the east on the coastal plain and *fuliginatus* to the northwest in the mountains. It seems best to refer the mice from 4 km. N Ciudad Valles to *B. t. taylori* on the basis of the average of external and cranial characters. Specimens from 6 mi. SW San Geronimo, Coahuila, also referred to *B. t. taylori*, resemble in color the mice from 4 km. N Ciudad Valles. When more specimens are obtained from the front range of the Sierra Madre Oriental, at lower altitudes, the manner in which these two subspecies intergrade with one another will be better understood. At present, populations from higher altitudes in the mountains seem to represent a dark subspecies; populations from the coastal plain represent a pale subspecies, and those from the lower slopes and high valleys seemingly are intergrades. *B. t. fuliginatus* occurs in a somewhat limited strip of chernozem soil (or suelos negros of Tamayo, 1949: Carta de Suelos). The populations occurring at lower altitudes on the coastal plain are on generally paler soils.

Specimens examined.--Total 39, all from the Republic of Mexico, as follows: SAN LUIS POTOSI: El Salto, 24 Mus. Nat. Hist., Louisiana State Univ., 7 Amer. Mus. Nat. Hist.; type locality, 8 (including the type).

Marginal records.--See specimens examined.

Baiomys taylori paulus (J. A. Allen)

Peromyscus paulus, J. A. Allen, Bull. Amer. Mus. Nat. Hist., 19:598, November 12, 1903; Elliot, Field Columb. Mus. Publ., 105(6): 136, July 1, 1905.

Baiomys taylori paulus, Miller, Bull. U. S. Nat. Mus., 79:137, December 31, 1912 (part); Miller, Bull. U. S. Nat. Mus., 128:317, April 29, 1924 (part); Ellerman, The Families and Genera of Living Rodents, 2:402, March 21, 1941 (part); Goldman, Smith, Miscl. Coll., 115:373, July 31, 1951 (part); Hall and Kelson, Univ. Kansas Publs., Mus. Nat. Hist., 26:367, December 15, 1952; Goodwin, Bull. Amer. Mus. Nat. Hist., 102:318, August 31, 1953; Miller and Kellogg, Bull. U. S. Nat. Mus., 205:511, March 3, 1955 (part); Packard, Proc. Biol. Soc. Washington, 71:17, April 11, 1958; Packard, Jour. Mamm., 40:146, February 20, 1959; Hall and Kelson, The Mammals of North America, 2:659, March 31, 1959 (part).

[*Peromyscus*] *paulus*, Elliot, Field Columb, Mus. Publ., 95(4):136, July 15, 1904.

Peromyscus taylori paulus, Osgood, N. Amer. Fauna, 28:255, April 17, 1909 (part).

Peromyscus musculus [*musculus*], Osgood, N. Amer. Fauna, 28:256, April 17, 1909 (part).

Baiomys taylori [= *paulus*], Twente and Baker, Jour. Mamm., 32:121, February 15, 1951.

Baiomys musculus musculus, Goldman, Smith. Miscl. Coll., 115:336, July 31, 1951 (part).

Baiomys taylori allex, Hall and Kelson, The Mammals of North America, 2:659, March 31, 1959 (part).

Type.--Adult male, skin and skull; No. 21165, American Museum of Natural History; Rio Sestin, Durango, Republic of Mexico; obtained on April 15, 1903, by J. H. Batty, original number 455.

Range.--Central Chihuahua south through Durango (west to eastern edge of Sierra Madre Occidental), to Zacatecas and Aguascalientes, thence west into northern and northwestern Jalisco, see Figure 11. Zonal range: Lower Sonoran, approximately the Chihuahua Desert Biotic Province of Goldman and Moore (1945:349). Occurs from 4000 feet 2 mi. ESE Tequila, Jalisco, up to 6700 feet 2 mi. W Minaca, Chihuahua.

Diagnosis.--Size medium to small for the species; dorsum Buffy Brown to fawn color; dorsal ground color of unworn pelage of adults varying from Buffy Brown in darkest series (especially those from higher altitudes) to Avellaneous with grayish overtones in palest series; worn pelage in mid-dorsal region of adults fawn to grayish; terminal parts of individual hairs buffy, gray basally; guard hairs on dorsum black-tipped, grayish basally; belly Light Gull Gray, distal half of hairs white, proximal half Neutral Gray; hairs in region of throat and chin white to base (some specimens with faint buffy overtones); forefeet dusky below, whitish above; hind feet whitish above, ventral surface whitish to dusky; dorsal and lateral vibrissae black, other vibrissae white. Average and extreme measurements of six adults from the type locality are as follows: total length, 109 (106-117); length of tail, 44.5 (43-48); length of body, 63 (57-69); length of hind foot, 13.1 (12.7-14.0); occipitonasal length, 17.5 (17.4-18.0); zygomatic breadth, 9.3 (9.1-9.5); postpalatal length, 6.6 (6.2-6.9); least interorbital breadth, 3.5 (3.4-3.6); length of incisive foramina, 3.8 (3.6-4.1); length of rostrum, 5.9 (5.7-6.0); breadth of braincase, 8.6 (8.5-8.8); depth of cranium, 6.6 (6.2-6.9); alveolar length of maxillary tooth-row, 3.2 (3.1-3.4); for photographs of the skull, see Plate 2*e* and Plate 4*f*.

Comparisons.--For comparisons with *B. t. allex*, *B. t. canutus*, *B. t. ater*, and *B. t. taylori*, see accounts of those subspecies. From *B. t. analogous*, *B. t. paulus* differs as follows: dorsal color paler having more reddish-brown than blackish-brown tones; venter whitish to buffy, instead of gray to light-gray; tail bicolored (not unicolored), usually having more hairs; hind feet white (not sooty) above. Cranially, *B. t. paulus* differs from *B. t. analogous* in: skull slightly smaller in all dimensions; maxillary part of zygoma narrowing and forming oblique angle rather than a near right angle with rostrum; anterior incisive foramina constricted posteriorly; tips of nasals truncate (less rounded).

Remarks.--J. A. Allen (1903:599) correctly pointed out that young specimens, in first pelage, were gray brown; young adults were darker and more varied with some blackish; adults and old adults were buffy to grayish. The change in color of pelage with increasing age is more pronounced in *paulus* than in other subspecies of *B. taylori*. Of two males collected on April 12, 1949, one, an adult, is buffy brown, and the other, an old adult with worn pelage, is grayish-brown. In mice in the earlier stages of adulthood, underfur of the dorsum is buffy at the tips and gray basally. With increased wear, the buffy tip is lost. Consequently, mice in the later stages of adulthood are grayish.

B. t. paulus intergrades with *ater* to the north in Chihuahua (see account of that subspecies), with *analogous* to the south in Jalisco, and with *allex* (see account of that subspecies) to the southwest in Nayarit and Jalisco. The zone of intergradation between *paulus* and *analogous* in Jalisco approximately borders the Rio Grande de Santiago from the western part of the State to the northwest shore of Lago de Chapala. Nineteen specimens from 2 mi. WNW Lagos de Moreno in northwest Jalisco seem to be intermediate between *paulus* and *analogous* in color, averaging slightly grayer than typical *paulus*. The series of 19 is referable to *paulus* on the basis of cranial characters.

A series of 34 specimens from 3 mi. W La Venta, Jalisco (referable to *paulus*), is indistinguishable in color of pelage from two series of *paulus* from 5 mi. N Durango, and from 8 mi. NE of Durango, except that the antiplantar surfaces of the hind feet are sooty as in *analogous*. Seemingly, features of color mentioned above as diagnostic of the two subspecies are either present or absent and there is no tendency toward intermediacy in color in the population from 3 mi. W La Venta.

The Rio Grande de Santiago may have acted in the past as a physical barrier reducing gene flow between *allex* and *paulus* and in separating completely the two populations for limited periods.

Specimens examined.--Total 176, all from the Republic of Mexico and distributed as follows: CHIHUAHUA: Rancho Sanignacio, 4 mi. S, 1 mi. W Santo Tomas, 1; El Rosario, 6700 ft., 1; 2 mi. W Minaca, 6900 ft., 11; Balleza, 1[50]. DURANGO: Rosario, 1[51]; type locality, 14[51] (including the type); *San Gabriel*, 2[51]; *Rancho Santuario*, 2[51]; 1 mi. N Chorro, 6450 ft., 1; *8 mi. NE Durango*, 6200 ft., 2; 5 mi. N Durango, 6400 ft., 2. ZACATECAS: Valparaiso, 6500 ft., 10[50]. AGUASCALIENTES: *18 mi. W, 2 mi. S Aguascalientes*, 6000 ft., 1; 16 mi. S Aguascalientes, 5[52]. JALISCO: 1 mi. NE Villa Hidalgo, 6500 ft., 1; 2 mi. WNW Lagos de Moreno, 6370 ft., 19; *2 mi. ESE Tequila*, 4000 ft., 11; *3 mi. W La Venta*, 33, 1[53]; *12 mi. W Guadalajara*, 3[54]; *Atemajac*, 12[50]; 4 mi. W Guadalajara, 5100 ft., 3; *2 mi. N, 1/2 mi. W Guadalajara*, 11; 2 mi. NW Magdalena, 4500 ft., 7[50]; *1 mi. N Tala*, 4400 ft., 3; 3 mi. W Tala, 4300 ft., 18.

Marginal records.--CHIHUAHUA: Rancho Sanignacio, 4 mi. S, 1 mi. W Santo Tomas; El Rosario; Balleza. DURANGO: Rosario, 6700 ft.; 1 mi. E Zarca (Blossom and Burt, 1942:1); 1 mi. N Chorro, 6450 ft. ZACATECAS: Valparaiso, 6500 ft. AGUASCALIENTES: 1 mi. N Chicalote (Blossom and Burt, 1942:4). JALISCO: 2 mi. WNW Lagos de Moreno, 6370 ft.; 4 mi. W Guadalajara, 5100 ft.; 3 mi. W Tala, 4300 ft.; 2 mi. NW Magdalena, 4500 ft. DURANGO: 5 mi. N Durango, 6400 ft.; type locality. CHIHUAHUA: 2 mi. W Minaca, 6900 ft.

[50] United States National Museum (Biol. Surv. Collections).

[51] American Museum of Natural History.

[52] Univ. Illinois, Mus. Nat. History.

[53] The Museum, Michigan State Univ.

[54] Univ. Michigan, Museum of Zoology.

Baiomys taylori subater (V. Bailey)

Peromyscus taylori subater, V. Bailey, N. Amer. Fauna, 25:102, October 24, 1905; Lyon and Osgood, Bull. U. S. Nat. Mus., 62:139, January 15, 1909; Osgood, N. Amer. Fauna, 28:255, April 17, 1909; Elliot, Check-List Mamm. N. Amer. Continent, West Indies and Neighboring Seas, Suppl., Amer. Mus. Nat. Hist, p. 44, January 8, 1917.

Baiomys taylori subater, Miller, Bull. U. S. Nat. Mus., 79:136, December 31, 1912; Miller, Bull. U. S. Nat. Mus., 128:317, April 29, 1924; Anthony, Field Book of North American Mammals, p. 348, 1928; Baker, Jour. Mamm., 21:223, May 14, 1940; Ellerman, The Families and Genera of Living Rodents, 2:402, March 21, 1941; Blair, Jour. Mamm., 22:378, November 14, 1941; Poole and Schantz, Bull. U. S. Nat. Mus., 178:259, March 6, 1942; Blair, Jour. Mamm., 23:196, May 14, 1942; Blair and Blossom, Contrib. Lab. Vert. Biol., Univ. Michigan, 40:1, March, 1948; Miller and Kellogg, Bull. U. S. Nat. Mus., 205:511, March 3, 1955; Hall and Kelson, The Mammals of North America, 2:659, March 31, 1959.

Baiomys taylori [= *subater*], Taylor and Davis, Texas Game, Fish and Oyster Comm. Bull., 27:56, August, 1947 (part).

Type.--Subadult female, skin and skull; No. 32616/44539 U. S. Nat. Mus. (Biol. Surv. Coll.); Bernard Creek, near Columbia, Brazoria County, Texas; obtained on February 25, 1892, by W. Lloyd, original number 1122.

Range.--Southeastern Texas, north of Matagorda Bay west to Lavaca County, north to Brazos and Walker counties thence east to Jefferson County, see Figure 11. Occurs from near sea level in Brazoria and Galveston counties, up to 500 feet in western part of range. Zonal range: Humid division of lower Austral (the western part of the Austroriparian Biotic Province of Dice, 1943:18-21).

Diagnosis.--Size medium to large for the species; mid-dorsal region Clove Brown (sooty in freshly captured specimens); some parts of mid-dorsal region all blackish; individual guard hairs of dorsum black-tipped, Deep Neutral Gray basally; underfur black-tipped with subterminal band of light buff, Neutral Gray at base; belly grayish-white, laterally Isabella Color; distal three-fourths of hairs in region of throat and chin white, proximal fourth light gray; in median region of belly distal half of individual hairs white, proximal half dark gray; vibrissae in most specimens black to base. Average and extreme cranial measurements of six adults from 7 mi. S La Belle are as follows: occipitonasal length, 18.9 (17.5-19.4); zygomatic breadth, 9.6 (9.1-9.9); postpalatal length, 6.8 (6.2-7.2); least interorbital breadth, 3.7 (3.4-3.9); length of incisive foramina, 4.0 (3.6-4.2); length of rostrum, 6.5 (6.1-6.8); breadth of braincase, 8.7 (8.3-8.9); depth of cranium, 6.7 (6.6-6.8); alveolar length of maxillary tooth-row, 3.1 (2.9-3.2). Average and extreme external measurements of four adults from Richmond are as follows: total length, 111.5 (108-118); length of tail vertebrae, 43.5 (41-47); length of body, 68 (67-71); length of hind foot, 14 (13-15); for photographs of the skull, see Plate 2*f*, and Plate 4*g*.

Comparisons.--Because *B. t. subater* intergrades only with *B. t. taylori* to the south and west, *subater* is compared only with *taylori*. Young adults of both subspecies in unworn pelage show best the colors that differentiate the two subspecies. Old adults of *subater* in worn pelage appear grayish, resembling *taylori*, and at that age, only certain cranial characters are of taxonomic use. Cranially, *subater* differs from *taylori* in: presphenoid not shaped like an hour-glass; parapterygoid processes thicker medially; interparietal diamond-shaped instead of elongated and compressed. Skull slightly larger in most measurements.

PLATE 1

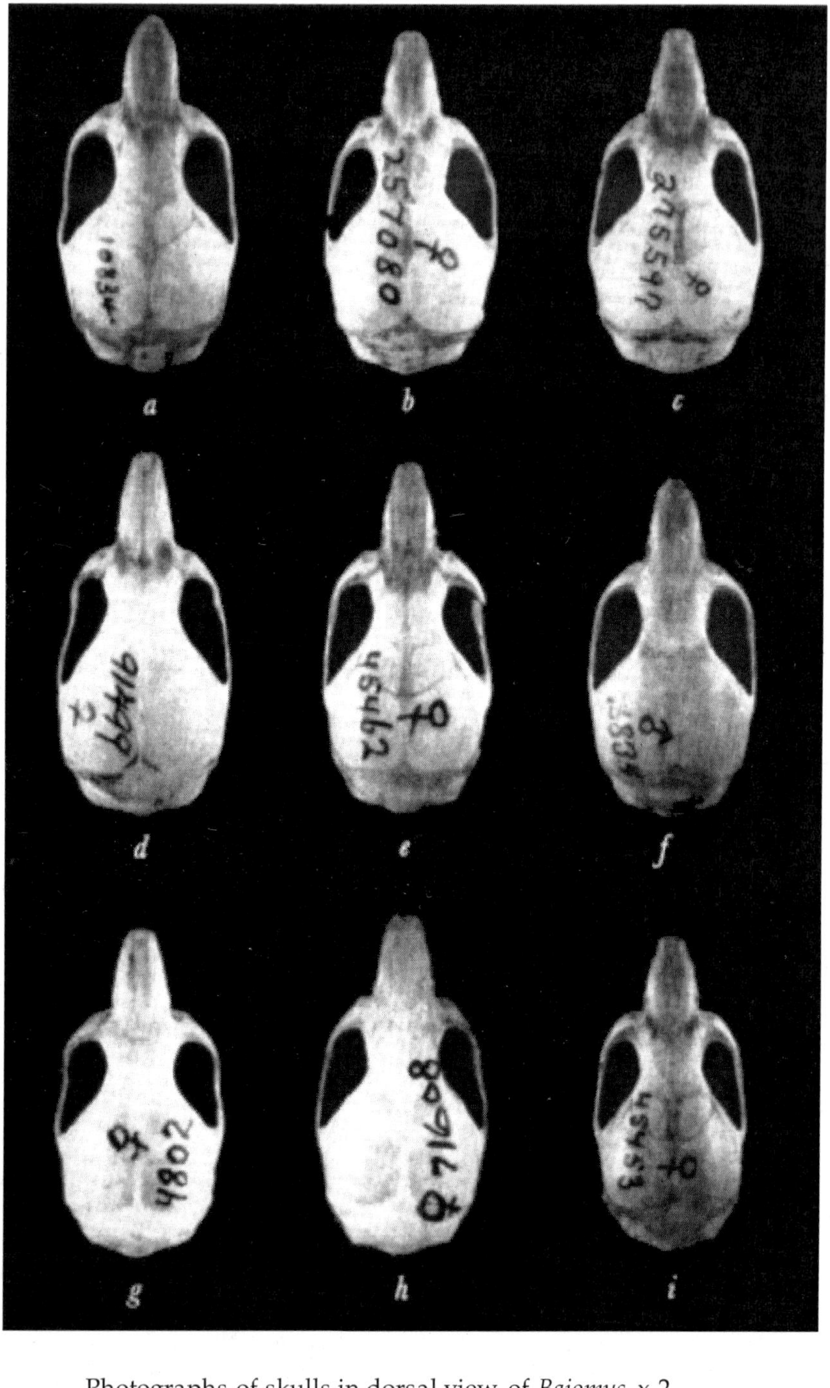

Photographs of skulls in dorsal view of *Baiomys*. x 2.

a. B. m. brunneus, ad., 10834, AMNH, Jalapa, Veracruz.
b. B. m. grisescens, ad., 257080, USNM, Comayabuela, Honduras.
c. B. m. handleyi, ad., 275597, USNM, Sacapulas, Guatemala.
d. B. m. infernatis, ad., 91499, MZUM, Teotitlan, Oaxaca.
e. B. m. musculus, ad., 45462, USNM, Colima, Colima.
f. B. m. nigrescens, ad., 76834, USNM, Comitan, Chiapas.
g. B. m. pallidus, ad., 4802, Texas A & M, Axochiapan, Morelos.
h. B. m. pullus, ad., 71608, KU, 8 mi. S Condega, Nicaragua.
i. B. t. allex, ad., 45453, USNM, Colima, Colima.

PLATE 2

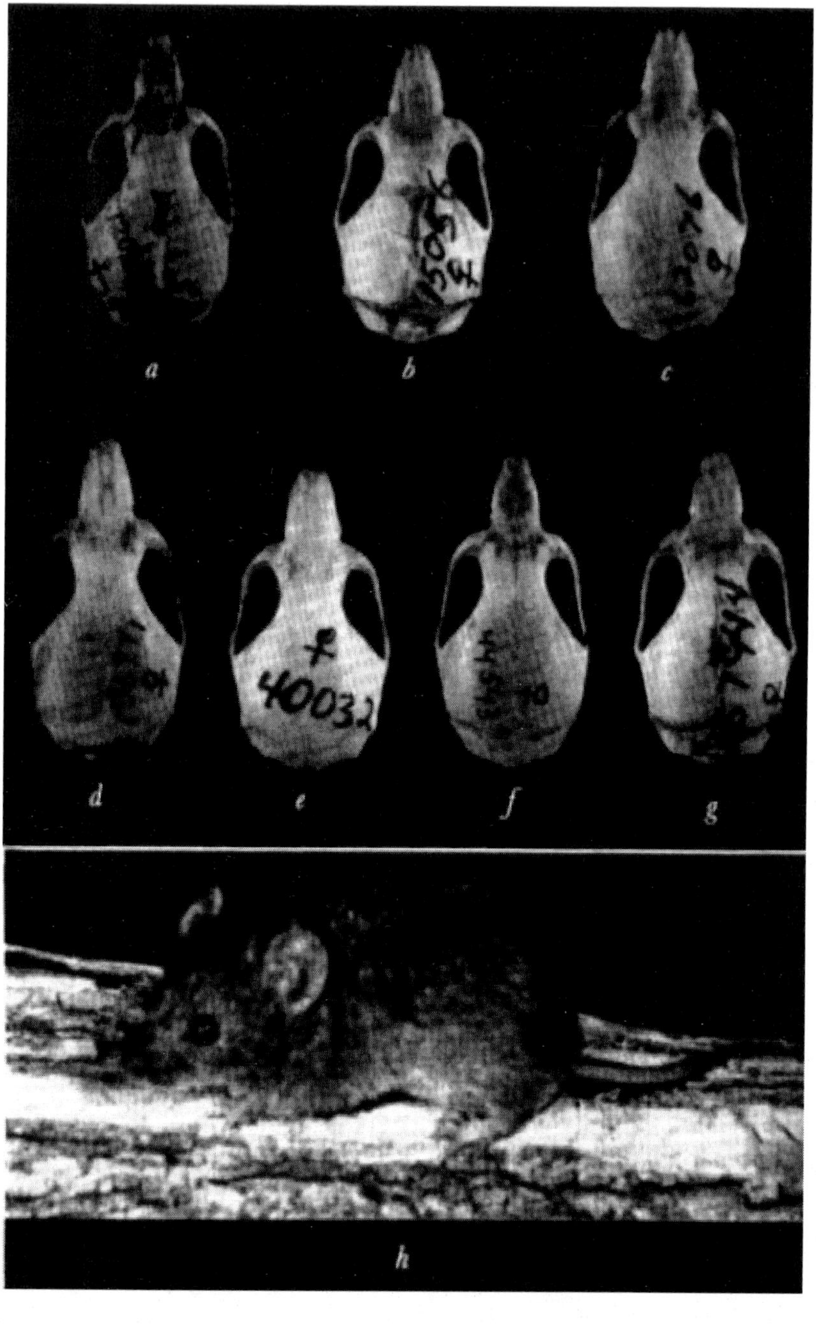

Photographs of skulls (*a-g*) in dorsal view of *Baiomys*. x 2.

a. *B. t. analogous*, ad., 120265, USNM, Zamora, Michoacan.
b. *B. t. ater*, ad., 15056, UI, 1½ mi. ENE Greaterville, Arizona.
c. *B. t. canutus*, ad., 62076, KU, 1 mi. S Pericos, Sinaloa.
d. *B. t. fuliginatus*, ad., 36771, KU, type locality.
e. *B. t. paulus*, ad., 40032, KU, 18 mi. W, 2 mi. S Aguascalientes.
f. *B. t. subater*, ad., 44543, USNM, type locality.
g. *B. t. taylori*, ad., 57944, KU, 5 mi. E San Antonio, Texas.
h. Photo. of captive *B. t. taylori*, 25 mi. E Austin, Texas. x 1.

PLATE 3

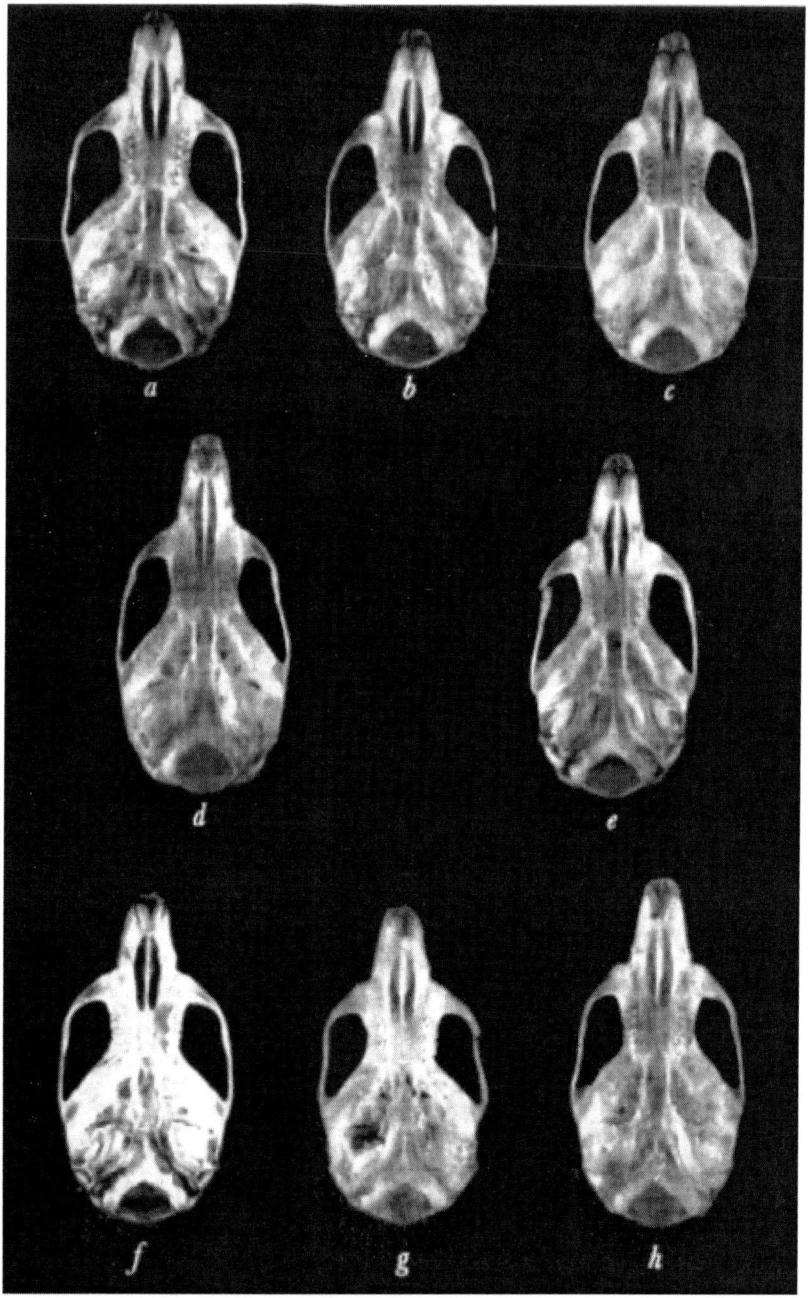

Photographs of skulls in ventral view of *Baiomys*. x 2.

a. B. m. brunneus, ad., 10834, AMNH, Jalapa, Veracruz.
b. B. m. grisescens, ad., 257080, USNM, Comayabuela, Honduras.
c. B. m. handleyi, ad., 275597, USNM, Sacapulas, Guatemala.
d. B. m. infernatis, ad., 91499, MZUM, Teotitlan, Oaxaca.
e. B. m. musculus, ad., 45462, USNM, Colima, Colima.
f. B. m. nigrescens, ad., 76834, USNM, Comitan, Chiapas.
g. B. m. pallidus, ad., 4802, Texas A & M, Axochiapan, Morelos.
h. B. m. pullus, ad., 71608, KU, 8 mi. S Condega, Nicaragua.

PLATE 4

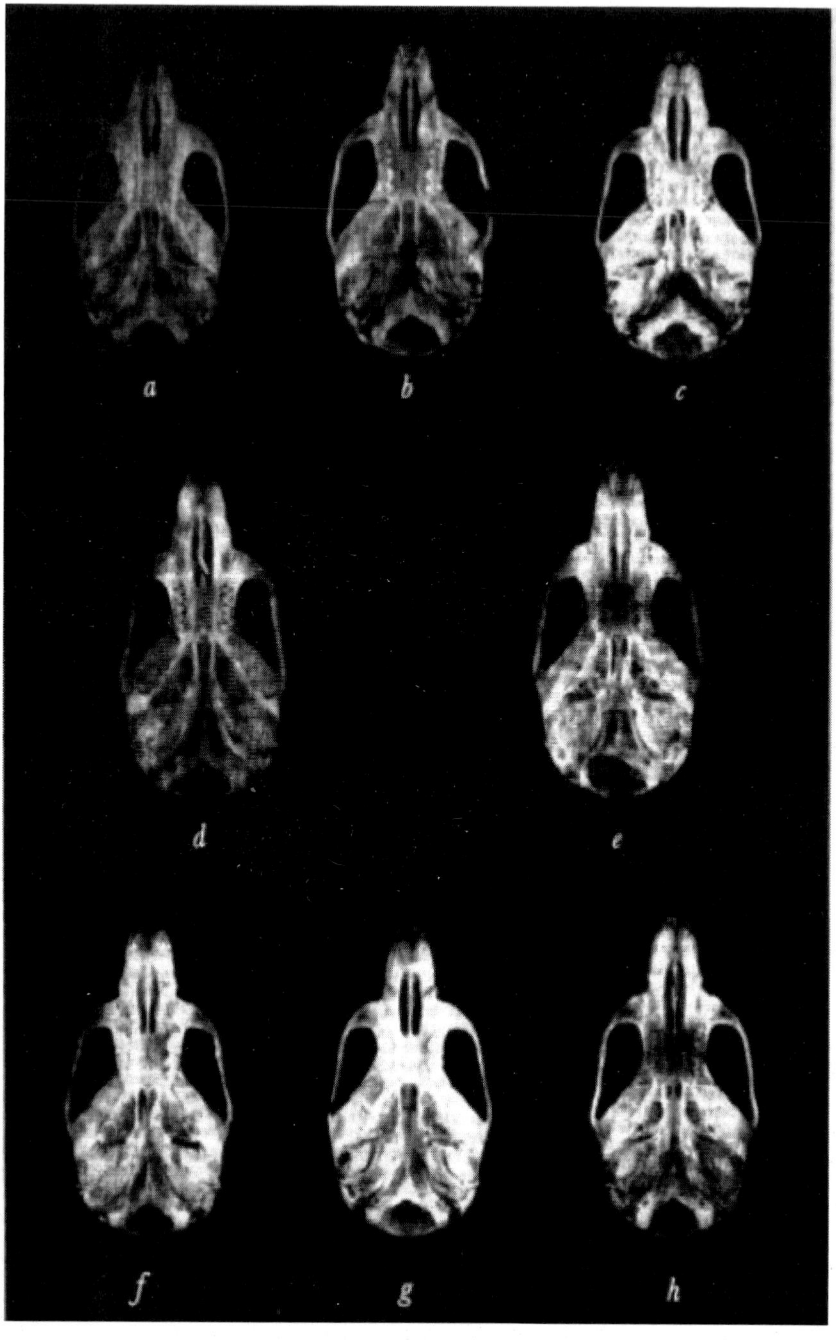

Photographs of skulls in ventral view of *Baiomys*. x 2.

a. B. t. allex, ad., 45453, USNM, Colima, Colima.
b. B. t. analogous, ad., 120265, USNM, Zamora, Michoacan.
c. B. t. ater, ad., 15056, UI, 1 mi. ENE Greaterville, Arizona.
d. B. t. canutus, ad., 62076, KU, 1 mi. S Pericos, Sinaloa
e. B. t. fuliginatus, ad., 36771, KU, type locality.
f. B. t. paulus, ad., 40032, KU, 18 mi. W, 2 mi. S Aguascalientes.
g. B. t. subater, ad., 44543, USNM, type locality.
h. B. t. taylori, ad., 57944, KU, 5 mi. E San Antonio, Texas.

Remarks.--This subspecies retains its chief diagnostic character, blackish mid-dorsal region, throughout nearly all parts of its range. Specimens from the general area of Matagorda Bay and Lavaca County grade into *taylori* in characters of color and crania. The Colorado and Brazos rivers seemingly serve as barriers reducing gene flow between *taylori* and *subater*. These rivers may well have been important factors in the origin and the limitation of these two seemingly closely-related subspecies.

Baiomys taylori subater is not differentiated in color of pelage and characters of crania from *B. t. taylori* to the same degree that *B. t. paulus* is differentiated from *B. t. analogous*, or that *B. t. taylori* is differentiated from several of the other subspecies of *Baiomys taylori*. *B. t. subater* probably is a more recent occupant of the area in which it now lives than is the case with any other one of the subspecies of *taylori*. Sufficient time probably has not elapsed to allow for formation of more distinctive phenotypic patterns.

Specimens examined.--Total 65, all from TEXAS and distributed as follows: *Brazos County*: 1/2 mi. NW College Station, 1[55]; *3 mi. W College Station, 1 mi. W Easterwood Airport*, 1[55]; *College Station*, 1[55]. *Walker County*: Huntsville, 1[55]. *Hardin County*: Sour Lake, 1[57]. *Jefferson County*: 7 mi. S Labelle, 10. *Harris County*: 6 mi. NE Crosby, 1[56]. *Colorado County*: *10 mi. N Eagle Lake*, 1[55]; *9 mi. N Eagle Lake*, 1[55]; 2 mi. W Eagle Lake, 1; *Eagle Lake*, 1[55], 5. *Fort Bend County*: Richmond, 4[57]. *Galveston County*: *Texas City*, 6[58]; Virginia Point, 1[57]. *Brazoria County*: *Austin Bayou near Alvin*, 2[57]; 14 mi. SSE Alvin, 2[59]; type locality, 7[57] (including the type). *Lavaca County*: 4 mi. W Hallettsville, 1[55]; *1 mi. SW Hallettsville*, 3[55]; *13.7 mi. SW Hallettsville*, 2[55]; 4 mi. NE Yoakum, 11.

Marginal records.--TEXAS: Huntsville; Sour Lake; 7 mi. S La Belle; Virginia Point; 14 mi. SSE Alvin; type locality; 4 mi. NE Yoakum; 4 mi. W Hallettsville; 1/2 mi. NW College Station.

[55] Texas A & M, Cooperative Wildlife Research Collection.

[56] Carnegie Museum.

[57] U. S. Nat. Museum (Biol. Surv. Coll.).

[58] Los Angeles County Museum.

[59] American Museum of Natural History.

Baiomys taylori taylori (Thomas) *Hesperomys* (*Vesperimus*) *taylori* Thomas, Ann. Mag. Nat. Hist., ser. 5, 19:66, January, 1887.

Baiomys taylori [*taylori*], Mearns, Bull. U. S. Nat. Mus., 56:381, April 13, 1907; Stickel and Stickel, Jour. Mamm., 30:141, May 23, 1949.

Baiomys taylori taylori, Miller, Bull. U. S. Nat. Mus., 79:136, December 31, 1912; Miller, Bull. U. S. Nat. Mus., 128:317, April 29, 1924; Anthony, Field Book of North American Mammals, p. 327, 1928; Ellerman, The Families and Genera of Living Rodents, 2:402, March 21, 1941; Taylor and Davis, Texas Game, Fish and Oyster Comm. Bull., 27:56, August, 1947 (part); Blair, Texas Jour. Sci., 2:104, March 31, 1950; Goldman, [Pg 652] Smith. Miscl. Coll., 115:373, 426, July 31, 1951; Baker, Univ. Kansas Publs., Mus. Nat. Hist., 5:212, December 15, 1951; Blair, Texas Jour. Sci., 4:242, June 30, 1952; Hooper, Occas. Papers, Univ. Michigan, Mus. Zool., 544:7, March 25, 1953; Dalquest, Louisiana State Univ. Studies (Biol. Sci. Ser.), 1:155, December 28, 1953 (part); Blair, Adv. in Genetics, 5:10, January 27, 1954; Miller and Kellogg, Bull. U. S. Nat. Mus., 205:511, March 3, 1955; Baker, Univ. Kansas Publs., Mus. Nat. Hist., 9:273, June 15, 1956; Packard, Proc. Biol. Soc. Washington, 71:17, April 11, 1958; Hall and Kelson, The Mammals of North America, 2:659, March 31, 1959 (part).

Cricetus (*Vesperimus*) *taylori*, Thomas, Proc. Zool. Soc. London, 68:446, November 20, 1888.

Sitomys taylori, Merriam, Proc. Biol. Soc. Washington, 7:170, September 29, 1892.

Sitomys (*Baiomys*) *taylori*, True, Proc. U. S. Nat. Mus., 16(972):758, February 7, 1894; J. A. Allen, Bull. Amer. Mus. Nat. Hist., 6:181, May 31, 1894.

S. [*itomys*] *taylori*, Rhoads, Proc. Acad. Nat. Sci. Philadelphia, 46:256, October, 1894.

Peromyscus (*Baiomys*) *taylori*, J. A. Allen, Bull. Amer. Mus. Nat. Hist., 8:65, April 22, 1896.

[*Peromyscus*] *taylori*, Trouessart, Cat. Mamm., 1:517, 1898.

Peromyscus taylori [*taylori*], Elliot, Field Columb. Mus. Publ., 105(4):135, July 1, 1905; V. Bailey, N. Amer. Fauna, 25:101, October 24, 1905; Elliot, Field Columb. Mus. Publ., 115(8):203, 1907; Osgood, N. Amer. Fauna, 28:253, April 17, 1909.

Type.--Adult male, skin and skull; No. 87.11.24.1, British Museum, Natural History; San Diego, Duval County, Texas; obtained by William Taylor.

Range.--North-central to southeastern Texas, excluding the coastal plain north of the region of Matagorda Bay, thence south into the southern part of Tamaulipas and west into Coahuila and Nuevo Leon, see Figure 11. Occurs from near sea level in Texas up to 1500 feet in Coahuila. Zonal range: mostly Lower Austral (in Mexico and southeastern half of Texas, the Tamaulipas Biotic Province of Goldman and Moore, 1945:349, and Blair, 1952:230).

Diagnosis.--Size medium for the species; dorsum grayish in freshly taken specimens to Hair Brown in preserved specimens; individual guard hairs of dorsum black-tipped, grayish basally, underfur black-tipped with a subterminal band of olive-buff; sides of body pale-grayish near venter, individual hairs buffy proximally, grayish basally; belly pale grayish, individual hairs white-tipped, Pale Neutral Gray basally; throat and chin colored as is belly; forefeet and hind feet sooty-gray dorsally, sparsely-haired ventrally, thus appearing flesh-colored; tail unicolored gray to sooty-gray. Average and extreme cranial measurements of 22 adults from 6 mi. SW San Geronimo, Coahuila, are as follows: occipitonasal length, 18.0 (17.4-19.0); zygomatic breadth, 9.6 (9.2-10.2); postpalatal length, 6.5 (5.9-7.1); least interorbital breadth, 3.6 (3.3-3.8); length of incisive foramina, 4.0 (3.6-4.3); length of rostrum, 6.1 (5.7-6.7); breadth of brain case, 8.8 (8.5-9.1); depth of cranium, 6.5 (6.0-7.0); alveolar length of maxillary tooth-row, 3.1 (3.0-3.3). Average and extreme external measurements of 19 adults from 6 mi. SW San Geronimo are as follows: total length, 102.2 (95-115); length of tail vertebrae, 39.4 (21-46); length of body, 62.8 (53-76); length of hind foot, 14.0 (12-15); length of ear from notch, 10.7 (10-12); for photographs of skull, see Plate 2*g*, and Plate 4*h*.

[Pg 653]

Comparisons.--For comparisons with *B. t. subater*, *B. t. analogous*, and *B. t. fuliginatus*, see accounts of those subspecies. From *B. t. paulus*, found to the southwest, *B. t. taylori* differs as follows: dorsum grayish rather than fawn-colored; hairs on dorsal parts of forefeet and hind feet sooty-gray (not white to white-brown); venter gray to Light Drab-Gray, rather than whitish with gray overtones; tail unicolored instead of bicolored; skull averaging slightly larger over-all; maxillary part of zygoma forms right angle with rostrum rather than obtuse angle; incisive foramina extending posteriorly to anterior plane of first upper molars instead of to a transverse plane at middle of right and left first upper molars; bullae less inflated; interorbital region broader relative to length of skull; rostrum sloping gently from frontonasal suture to anterior tip of nasals rather than declining abruptly from frontonasal suture to anterior tip of nasals.

Remarks.--The geographic range of *taylori* is relatively large, and the subspecies is locally variable. Nevertheless, none of the external and cranial measurements of specimens assigned to this subspecies differs significantly from the corresponding measurements of material from the type locality and adjacent areas in southeastern Texas. In southeastern Texas, south of the Guadalupe River, south to the coastal plain of Tamaulipas, this subspecies differs in color (being paler) from *B. t. subater* with which *taylori* might be confused. The foothills of the Sierra Madre Oriental in western Tamaulipas, north through Nuevo Leon and Coahuila, seem to mark the southwestern limit of the range assignable to *taylori*.

On December 27, 1958, a specimen, KU 81552, was obtained 3 mi. N Bowie, Montague County, Texas. This record station extends the known range of *B. taylori* 65 miles northward from the previous northernmost locality, listed by Hunsaker, Raun, and Swindells (1959:447). Two specimens, KU 81553 and 81554, were collected by the author 2 mi. NE Cedar Hill, Dallas County, Texas, on October 31, 1958. These two specimens, plus the single specimen from Bowie County are all paler with more buffy bellies than either *B. t. taylori* or *B. t. subater*. They may represent an incipient subspecies. I tentatively assign them to *B. t. taylori* because of the pale rather than dark (like *B. t. subater*) pelage. Additional specimens are needed from these areas and from the hiatus between the ranges of *B. t. taylori* and *B. t. subater* the better to understand the manner in which these two subspecies intergrade.

Among named subspecies of *Baiomys taylori*, *B. t. taylori* most closely resembles *B. t. subater* to the north in Texas. Nine specimens examined from Yoakum are intergrades between *taylori* and *subater*. These specimens have the sooty dorsal color of *subater*, but ventrally are inseparable from topotypes of *taylori*. In length of body and tail, specimens from Yoakum are like *subater*, but in length of hind foot, they are intermediate between the two subspecies. Cranially, they are like *subater*. When all characters are considered, the specimens are best referred to *subater*. Bailey (1905:103) suggested that specimens from the southern part of the range, which he ascribed to *subater*, tended to a more grayish color than topotypes of *subater*, therefore, grading into *taylori*. The zone of intergradation runs from Matagorda Bay northwest through Lavaca County, thence north to the Colorado River, and closely follows the boundary between the Lower Austral and Humid Division of Lower Austral Life-zone as plotted by Bailey (*loc. cit.*). Findley (1955:44) pointed out that where two life-zones meet, the resulting populations of shrews are mostly intergrades. Such is the case between these two subspecies of *Baiomys taylori* in an area where life-zones might seem less important than in the mountainous west.

In the southern part of the range of *taylori*, intergradation occurs between *B. t. taylori* in western Tamaulipas and *B. t. fuliginatus* in the mountains of San Luis Potosi.

Dalquest (1953:156) found no indication of intergradation between the two species, *B. taylori* and *B. musculus*, in San Luis Potosi. After examination of specimens from San Luis Potosi, I am in agreement that they are all referable to the species *taylori*.

Specimens examined.--Total 435. TEXAS: *Montague County*: 3 mi. N Bowie, 1. *Dallas County*: 2 mi. NE Cedar Hill, 2. *Travis County*: 8 mi. NW Austin, 2[60]; *Austin*, 2[60]; *4 mi. E Austin*, 4[60]; *5 mi. E Austin*, 3[60]; *6 mi. E Austin*, 16[60], 1; *7 mi. E Austin*, 1[60]; *15 mi. E Austin*, 1[60]; *4 mi. S Austin*, 1[60]. *Bastrop County*: 25 mi. E Austin, 2. *Kendall County*: Boerne, 1[61]. *Bexar County*: *1 mi. N Randolph Field*, 3[64]; *5 mi. ENE (on U. S. Highway 81) San Antonio*, 1; *3 mi. NE San Antonio*, 1; San Antonio, 26[61], 11[62], 1[63]; *5 mi. E San Antonio*, 11; *4 1/2 mi. E Sayers*, 3. *Gonzales County*: 7 mi. S Luling, 2[60]. *Wilson County: 4 mi. W LaVernia*, 3; 12 mi. W Floresville, 1. *Atascosa County*: 9 mi. SW Somerset, 1. *Goliad County*: 8 mi. NE Goliad, 1[60]. *Bee County*: Beeville, 1[61]. *Aransas County*: Aransas (Wildlife) Refuge, 1[65]; *5 mi. E Copana Bay*, 1[65]; *4.6 mi. NE Rockport*, 5[60]; *4.5 mi. NW Rockport*, 2[60]; 3 mi. N, 2 mi. E Rockport, 4; *Rockport*, 1[60], 1[61], 1[63]; *1 1/2 mi. SW Rockport*, 1[60]; *2 mi. SW Rockport*, 2[60]; *13.4 mi. SW Rockport*, 1[60]; *14 mi. SW Rockport*, 1. *San Patricio County*: Welder Wildlife Refuge, 7. *Duval County*: type locality, 2[61], 1[66]. *Nueces County*: Corpus Christi (south Nueces Bay), 1[64] (Cleveland Mus. Coll.). *Kleberg County:* 2 mi. S Riviera, 3[65]. *Brooks County*: 3 mi. S Falfurrias, 2[65]. *Hidalgo County*: 6 mi. S McAllen, 17[60]. *Willacy County*: 28 mi. E Raymondville, 10[65]. *Cameron County*: Brownsville, 31[61], 23[62], 5[64]. COAHUILA: 6 mi. SW San Geronimo, 32. NUEVO LEON: Santa Catarina, 1[61]; 14 mi. N Monterrey, 1950 ft., 2[67]; Monterrey, 1[61]; 20 km. N General Teran, 3[64]. TAMAULIPAS: *Near Headwaters Rio Sabinas, 8 km. W, 10 km. N El Encino*, 400 ft., 1; Camargo, 5[61]; Charco Escondido, 20 mi. S Reynosa, 3[67]; Matomoras, 5[61]; *Ejido Santa Isabel, 2 km. W Inter-American Highway*, 2000 ft., 7; Hidaglo, 7[61]; *Hda. Station Engracia*, 4[63]; 4 mi. N La Pesca, 1; 29 mi. N Ciudad Victoria, 1[67]; Ciudad Victoria, 6[61], 3; Jaumave, 2400 ft., 6[64], 10; Sierra de Tamaulipas, 3[64]; *25 mi. N El Mante, 3 km. W Inter-American Highway (on Rancho Pano Ayuctle)*, 300 ft., 4; *6 mi. N Gomez Farias (on Rancho Pano Ayuctle)*, 1; *5 mi. NE Gomez Farias*, 12[64], 1[62]; 70 km. (by highway) S Ciudad Victoria, 2 km. W El Carrizo, 5[62], 2; Antigua Morelos, 5[64]; *6 mi. N, 6 mi. W Altamira*, 31; *5 mi. N, 5 mi. W Altamira*, 4; *Alta Mira (Altamira)*, 2[61]; 1 mi. S Altamira, 6; *10 mi. NW Tampico*, 1. SAN LUIS POTOSI: Ebano, 5[68]; *4 km. NE Ciudad Valles*, 1; Ciudad Valles, 1; *3 km. W Tamuin*, 1[68]; *Tamuin*, 6[68]; *Pujal*, 300 m., 1[64]. VERACRUZ: Tampico Alto, 50 ft., 1; Potrero Llano, 350 ft., 1; Ozulama, 2; Cerro Azul, 350 ft., 1.

Marginal Records.--TEXAS: 3 mi. N Bowie; 2 mi. NE Cedar Hill; 25 mi. E Austin; 7 mi. S Luling; 8 mi. NE Goliad; Aransas (Wildlife) Refuge; 3 mi. N, 2 mi. E Rockport; Corpus Christi (South Nueces Bay); 2 mi. S Riviera; 28 mi. E Raymondville; Brownsville. TAMAULIPAS: Matomores; 4 mi. N La Pesca; 1 mi. S Altamira. VERACRUZ: Tampico Alto; Ozulama; Cerro Azul; Potrero Llano. SAN LUIS POTOSI: Ciudad Valles. TAMAULIPAS: Antigua Morelos; 70 km. S Ciudad Victoria, 2 km. W El Carrizo; Jaumave; Hidalgo. NUEVO LEON: 20 km. N General Teran; Santa Catarina. COAHUILA: 6 mi. SW San Geronimo. TEXAS: 9 mi. SW Somerset; Boerne; 8 mi. NW Austin.

[60] Coll. University of Texas.

[61] U. S. Nat. Museum (Biol. Surv. Coll.).

[62] American Museum of Natural History.

[63] Chicago Natural History Museum.

[64] Univ. Michigan, Museum of Zoology.

[65] Texas A & M Coop. Wildlife Res. Coll

.

[66] Carnegie Museum.

[67] Univ. California, Mus. Vert. Zool.

[68] Museum of Natural History, Louisiana State University.

EVOLUTION AND SPECIATION

The history of the genus dates back to the early late Pliocene, but morphological change since then has been slight insofar as can be judged from lower jaws. *Baiomys* seems to have been relatively conservative also in types of habitat occupied.

According to Wilson (1937:59), the late Pliocene was a time of decided expansion of myomorph rodents, more particularly cricetines. Furthermore, at this time, the climate in the interior basin of southwestern North America presumably was becoming arid, if we can judge from the spread of elements of the Madro-Tertiary flora. Axelrod (1950:266) points out that the drier, continental climate initiated in the early Tertiary probably had its culmination in middle Pliocene time. Some floras of early late Pliocene of the southwestern United States reflect a climate slightly cooler and more moist than the climates of the middle Pliocene. However, late Pliocene times reflect an arid climate. The flora of the southwestern interior basin of North America in early late to late Pliocene was intermediate between the previous grassland floras of the middle Pliocene and the savannah flora of upper Pliocene. Axelrod (*loc. cit.*) suggests that this intermediate flora of the interior basin of southwestern North America resulted from the folding of the Cascades and uplifting of the Sierra Nevada and Peninsular ranges to the south. The development of these mountains produced greater aridity to the lee of the mountains, thus accounting for the grassland-savannah flora. Pygmy mice probably originated in that time, I judge in Mexico, and moved northward and southward in a grassland-savannah habitat that seemingly existed as far north as what is now Meade County, Kansas (where the Sawrock fauna lived). Further evidence for occupancy of a grassland-savannah habitat by ancestral pygmy mice stems from the distribution of the living species, *B. taylori*, that at present occupies territory adjacent to parts of the Sonoran and Chihuahuan deserts. *B. taylori* seems to be morphologically more specialized for life in an arid grassland than was *B. sawrockensis*.

The geographic range of ancestral pygmy mice possibly extended farther south in late Pliocene time than the range of *B. musculus* does now. Anyhow, *B. sawrockensis* of the early late Pliocene dwelt in a more mesic type of habitat than *B. musculus* does, and such habitat may have existed from the Pacific lowlands of Central America to the Caribbean lowlands of northern South America (see Duellman, 1958:136, and Dunn,

1940:156) during late Pliocene times. An ancestral stock of hesperomine mice, not greatly different from *Baiomys*, may have emigrated from the North American continent into South America across the continuous land connection, which Simpson (1950:395) suggests was formed in the Chapadmalalan age (= Blancan age of North American terminology). The length of time of interchange of genes between northern and southern populations of mice across the Central American land connection probably was brief. Duellman (*op. cit.*:129) pointed out that once the Panamanian portal was closed, the warm counter equatorial current, El Nino, combined with the uplifting of the Andes, began to produce heavy rain forests in Central America and northern South America in late Pliocene or early Pleistocene times. These forests presumably isolated the stock in North America from that in South America where the latter probably evolved rapidly into kinds that differed from one another and from *Baiomys* in shape of body, type of pelage, and shape of skull. Internal structures such as hyoid apparatus, auditory ossicles, and baculum remained almost unchanged, as for example in *Calomys* now living in South America. The present resemblance in internal morphological features between it and *Baiomys*, I judge, reflects taxonomic relationships more accurately than do shape and conformation of body and skull that seem to respond more rapidly to external environmental changes. The cranial characters distinguishing *Baiomys musculus* from *Calomys laucha* are as follows: posterior lacerate foramina between second, rather than first, upper molars; parapterygoid fossa shallower; mesopterygoid fossa as wide or wider, instead of narrower, than parapterygoid processes; burr for attachment of superficial masseter muscle hypertrophied instead of well-developed. In other cranial characters studied, the two genera closely resemble each other. Such similarities of crania between *Calomys* and *Baiomys* may reflect convergence, but the total of internal and external morphological characters shared, I think reflects true relationships.

Peromyscus has a large number of living and extinct species and exhibits a wide range of morphological variation, whereas *Baiomys* has a small number (7) of species and exhibits a narrow range of morphological variation. The small number of known species of pygmy mice suggests their conservatism in elaboration of morphological characters. Possibly this is because the habitat, or even the ecological niche, occupied in geological time by these mice was restricted, geographically and in kind. If the habitat of the pygmy mice oscillated between savannah and arid grassland, then an hypothesis can be made possibly accounting for the origin of species of these mice. My idea is that the geographical distribution of *Baiomys* today reflects a predilection on the part of these mice for a relatively uniform warm climate. Therefore, in the past, in times of warmer continental climate, these mice moved toward favorable habitat northward from an area in central and northern Mexico. In cooler periods, the mice moved southward as habitats to the north became unfavorable.

Dr. W. B. Davis (*in. litt.*) informs me that *B. taylori* was uncommon in Brazos County, Texas, approximately 15 years ago, and suggests that the abundance there now of this mouse and my taking it in 1958 northward nearly to the southern border of Oklahoma reflects a definite movement northward. Movement in the same direction in late years has been suggested for the nine-banded armadillo and the hispid cotton rat (Hall, 1959:373) that are associated with warm climates to the south. These movements possibly reflect only minor fluctuations of climate, but in a long period of warmth movements northward would be expected to be pronounced and extensive.

Extinct species of *Baiomys* may have originated as a result of extension northward of the geographical range and subsequent retreat southward of the northern populations, as follows: (1) the range of the genus moved northward in a warm period; (2) in cooler times, most of the mice in the north disappeared and only isolated colonies remained in small patches of remaining habitat still favorable to the mice; (3) the small populations of isolated pygmy mice after a time changed through mutations, recombinations and subsequent selection to a degree that prevented crossbreeding once populations from the south again moved northward and came in contact with previously isolated stocks; (4) then competition caused further divergence in morphological characters. Such an hypothesis would account for the morphological differences between the extinct *B. kolbi* and *B. rexroadi*. The extinct *B. brachygnathus*, presumably a dweller of a xerophytic grassland, may have had its origin from a *B. minimus*-like stock in the manner outlined.

FORMATION OF THE RECENT SPECIES

The morphological difference between the extinct *B. minimus* and the living *B. musculus* is not great, and musculus seems to be the product of the *B. sawrockensis-B. minimus* line of development. Morphological characters of the parental stock of the two living species, *musculus* and *taylori*, may have been intermediate between those of *B. minimus* and those of *B. musculus*. The principal part of the range of *Baiomys* today is in Mexico, and probably was there through much of Pleistocene time. Extension northward of the species and retreat southward of those northern populations of pygmy mice would not only have left isolated populations in the north, but would have allowed the mice that retreated south to share a common gene pool. Therefore, populations of pygmy mice occurring to the south in central Mexico might be expected to maintain a relatively high degree of heterozygosity in morphological and behavioral characters. The occurrence of any physical or biotic barrier that would have separated this homogeneous group would be conducive to speciation. There is evidence that a barrier occurred in the Pleistocene in central Mexico sufficient to separate the supposed interbreeding, relatively homogeneous populations of pygmy mice. According to Sears (1955:529) and De Terra *et al.* (1949:51), parts of the higher regions in the Valley of Mexico, and the transverse volcanic zone in central Mexico were glaciated. On the mountain Ixtaccihuatl, De Terra (*op. cit.*:52) found evidence of four marked advances of ice, from oldest to youngest, as follows: Salto, ice advanced to 3100 meters; Xopano, ice at 3200-3300 meters; Trancas, ice to 3400 meters; Ayolotepito, ice to 4350 meters. The Salto advance is correlated by De Terra (*loc. cit.*) with the Iowan glacial period. The advance of ice down the mountain sides in the transverse volcanic zone was accompanied by cool moist climates or pluvial periods. Such climates probably altered habitat formerly suitable for *Baiomys*. There is no record of *Baiomys* known to me exceeding 8000 feet in elevation, although the lower edge of the ice on Ixtaccihuatl is at approximately 15,300 feet (4600 meters, Sears, *loc. cit.*). Presumably, the advance of ice down the mountains forced the pygmy mice to move to lower altitudes. Pluvial conditions possibly rendered the habitat even at lower altitudes uninhabitable for the mice, with the result that none continued to live in the transverse volcanic zone, but only north and south thereof. Long-continued separation of these northern and southern segments allowed species formation to occur. As climatic and habitat conditions became more favorable in central Mexico, the two species moved back toward each other, and eventually their geographic ranges overlapped.

An analysis of external and cranial characters of pygmy mice (see Figure 12) reveals that both species are essentially largest to the north and smallest to the south. There are exceptions to this cline in both species. For example, *B. taylori analogous* is a large subspecies; it lives allopatrically in the southern part of the range of the species. *B. musculus pallidus* is not the largest subspecies; it lives allopatrically in the northern part of the range of the species. In west-central Mexico, where the two species are sympatric, *B. taylori* is smaller than elsewhere and *B. musculus* is larger than elsewhere. *B. t. analogous* lives in the mountains of the transverse volcanic zone in central Mexico. Its large size may be a result of the cooler climate in the mountains. *B. t. allex*, the smallest subspecies, lives sympatrically with *B. musculus musculus* at lower elevations in west-central Mexico. The small size of *allex* could be a result of the warmer climate of the lower elevations. *B. m. pallidus*, at lower elevations in southern Oaxaca, is smaller than other subspecies of *musculus* to the south at higher elevations. *B. m. musculus* lives at low elevations along the coast of west-central Mexico. Unlike *B. m. pallidus*, *B. m. musculus* is large at lower elevations. It occurs sympatrically with *B. t. allex*. It is my idea that during the period of separation, when the two species were evolving, larger subspecies evolved to the north or at higher altitudes where climates were cooler; smaller subspecies evolved to the south or at lower elevations; the two cognate species, *musculus* and *taylori*, made contact at lower elevations where individuals of *taylori* may have been smallest, but individuals of *musculus* were not the largest of the species. The differences, therefore, between the two species in their initial contact probably were slight. Hybrids, if they occurred, were probably inviable, sterile, or ill-suited for occupancy of the habitat of either of the parental stocks. The occurrence of hybrids, therefore, would result in what geneticists call "gamete wastage," and any further divergence in the parental stock, either in external characters (size and shape of body and head), or behavior, useful in recognition of species, would be favored by natural selection (see Dobzhansky, 1951:225; and Koopman, 1950:147). The two species seem to have diverged more in external characters where they occur together than in areas where they live separately (see Figure 12). The two species could be confused if a sample of adults of *taylori* from 7 mi. S La Belle, Jefferson County, Texas, were compared to a sample of adults of *musculus* from Tehuantepec, Oaxaca (see Figure 12). No confusion in species identity would arise, however, if a sample of adults was taken from the area where the two species live together (see Figure 12). Brown and Wilson (1956:49) pointed out that where two closely related species occur together, characters (morphological, ecological, physiological, or behavioral) of each species are easily distinguished. However, where the two species are allopatric, the two closely related species so resemble one another that the species are not easily distinguished. This phenomenon has been called "character displacement" by Brown and Wilson (*loc. cit.*).

In the area where the two species of pygmy mice occur together, there seems to be a disparity in numbers between them. Hooper (1952a:91) has recorded the collection of both *B. musculus* and *B. taylori* in a single trap line. A series of pygmy mice collected from San Gabriel, Jalisco, contained one *taylori* and 33 *musculus*; another sample from La Resolana, Jalisco, had a ratio of 25 *taylori* to 6 *musculus*. The disparity in numbers where the two species occur together has been further substantiated by collections of the University of Kansas. Possibly this disparity in numbers is a result of interspecific competition. Hooper (*op. cit.*:90) pointed out that where the range of *B. musculus* (typical of arid tropical lowlands) meets that of *B. taylori* (typical of arid temperate highlands), the two geographic ranges interdigitate with parts of the range of *musculus* extending into the highlands and parts of the range of *taylori* extending into the lowlands. In the lowlands, *musculus* may be better adapted to environmental conditions and, therefore, more successful in competition with *taylori* for available habitat. The reverse situation may exist in the highlands. Also, the fact that *musculus* is more of a diurnal animal than is *taylori* may account for the difference in numbers of individuals of the two species taken in trap lines. Many collectors set their traps in late afternoon or evening and retrieve them in early morning. Such a schedule might not yield many *musculus*. If interspecific competition does occur in the area where the two species occur, any change in habits or microhabitat by either species that would reduce this competition would be favored by natural selection (see Mayr, 1949:518; Lack, 1944:262-263; and Brown, 1958:154-155). Brown (*op. cit.*:154), as I understand him, pointed out (taking account of Gause's principle) that when two species having similar ecological valences move into the same niche in the same locality, one of three things must eventually happen: (*a*) the two species occupy different geographic ranges; (*b*) they compete and one is eventually eliminated; (*c*) the two species, because of differentiation or specialization, exploit different aspects of the niche. In *Baiomys*, (*c*) seems to apply. Natural selection probably would favor a continuation of diurnal activity in *musculus* and nocturnal activity in *taylori*, thereby preventing frequent meeting of the two species.

AREAS OF PRESENT DIFFERENTIATION

In both species of *Baiomys*, the most distinct subspecies, *B. t. allex* and *B. m. musculus*, occur in the area where the two species are sympatric. Seven subspecies, or 44 per cent, occur either in or adjacent to the transverse volcanic zone. This area is the major area of active differentiation. Incipient subspecies are also evident in these areas. A secondary area of differentiation is indicated within the range of *B. musculus* in Guatemala, El Salvador and Honduras. Three subspecies occur in this area (*grisescens*, *handleyi* and *nigrescens*) and incipient subspeciation is in evidence there.

ZOOGEOGRAPHIC POSITION

Hooper (1949:25) regards *Baiomys* as a member of the rodent fauna of the arid, western Sonoran region, whereas Hershkovitz (1958:609) suggests that *Baiomys* is a nearctic-neotropical varicant (a kind that occurs in contiguous zoogeographic regions without our knowing in which region the taxon originated). The findings from my study do not contradict either of the above suggestions. Because of the close resemblance of *Baiomys* to certain hesperomine mice of South America, it is postulated that *Baiomys*, in more primitive form than now, occurred farther south in past times than it does now. Fossils show that primitive stocks of the genus in late Pliocene or early Pleistocene times occurred also north of the present range of the genus. The belt in west-central Mexico between nearctic and neotropical regions is the current center of distribution of the genus and probably has been for a considerable time.

FIG. 12. Averages of the occipitonasal lengths of skulls of adults at 19 localities of occurrence (solid symbols) of *Baiomys taylori*, and at 17

localities of occurrence (open symbols) of *Baiomys musculus*. Note that the occipitonasal length decreases from north to south in each of the two species, and that in the region where the two species occur together, west-central Mexico, *B. taylori* is smallest and *B. musculus* is largest. Average, extremes, number of specimens averaged (in italic type), and name of locality, from north to south for each species, are as follows:

Baiomys taylori

18.0 (17.5-18.6) *15*, 9½ mi. W New Mexico state line, Ariz.
18.9 (18.2-19.4) *6*, 7 mi. S. La Belle, Jefferson Co., Texas.
18.2 (17.8-18.5) *10*, San Antonio, Bexar Co., Texas.
18.2 (18.0-18.5) *5*, 2 mi. W Minaca, Chihuahua.
18.0 (17.6-19.0) *22*, 6 mi. SW San Geronimo, Coahuila.
18.2 (18.1-18.3) *3*, Ciudad Obregon, Sonora.
18.1 (17.4-18.5) *5*, vic. (see p. 649) Durango, Durango.
18.1 (17.5-18.5) *9*, Jaumave, Tamaulipas.
18.2 (17.7-18.9) *19*, 15 mi. N Rosario Chele, Sinaloa.
17.9 (17.4-18.3) *27*, vic. (see p. 655) Altamira, Tamaulipas.
18.3 (17.9-18.7) *9*, Valparaiso, Zacatecas.
18.1 (18.1-18.2) *4*, Ciudad del Maiz, San Luis Potosi.
18.6 (18.3-18.9) *8*, Tepic, Nayarit.
18.0 (17.7-18.4) *18*, 4 mi. N, 5 mi. W Leon, Guanajuato.
18.1 (17.5-18.9) *28*, 6 mi. E Queretaro, Queretaro.
17.7 (17.1-18.1) *17*, 1 mi. SSE Ameca, Jalisco.
17.3 (16.8-17.9) *10*, 2 mi. SSE Autlan, Jalisco.
18.0 (17.5-18.6) *10*, 1 mi. S, 11 mi. W Zamora, Michoacan.
17.6 (17.4-18.2) *8*, Colima, Colima.

Baiomys musculus

20.2 (19.9-20.3) *6*, vic. (see p. 622) Ameca, Jalisco.
20.2 (19.9-20.3) *6*, 2 mi. SSE Autlan, Jalisco.
19.6 (19.2-20.1) *6*, Jalapa, Veracruz.
20.3 (19.7-20.9) *9*, Colima, Colima.
19.5 (19.0-20.0) *10*, Cerro Gordo, Veracruz.
19.8 (19.4-20.3) *6*, 6 mi. S Izucar de Matemores, Puebla.
20.0 (18.8-20.5) *7*, Teotitlan, Oaxaca.
20.1 (19.7-20.7) *7*, 1 km. NW Chapa, Guerrero.

19.9 (19.4-20.4) *8*, 5 mi. ESE Tecpan, Guerrero.
19.5 (19.1-20.1) *22*, 3 mi. ESE Oaxaca, Oaxaca.
19.5 (19.1-19.9) *11*, Valley of Comitan, Chiapas.
18.9 (18.2-20.1) *17*, Tehuantepec, Oaxaca.
18.9 (18.4-19.7) *15*, 6 mi. NW Tonala, Chiapas.
19.1 (18.8-20.4) *10*, 1 mi. S Rabinal, Guatemala.
19.7 (18.8-20.4) *10*, Lake Amatitlan, Guatemala.
19.2 (18.4-19.8) *26*, vic. (see p. 625) San Salvador, El Salvador.
19.3 (18.9-19.9) *24*, 8 mi. S Condega, Esteli, Nicaragua.

CONCLUSIONS

1. Two Recent species, each polytypic with eight subspecies, and five fossil species are recognized.

2. The phyletic trends in the genus *Baiomys* have been from an ancestral stock that possessed relatively brachydont teeth having raised cingular ridges and orthodont to proodont incisors, to species having hypsodont teeth with reduced cingular ridges and retrodont incisors.

3. Reduction of cingular ridges in pygmy mice is associated with an existence in open grassland (more xeric than mesic), whereas, the presence of cingular ridges is associated with an existence in a savannah habitat (more mesic than xeric).

4. Shifts of geographical range of populations of pygmy mice at and near the periphery of their geographic range may account for the differentiation of the extinct species.

5. The two living species, *B. musculus* and *B. taylori*, are seemingly derived from a common ancestor that in morphological structure was intermediate between *B. minimus* and *B. musculus*.

6. The living species of pygmy mice resulted from a geographic separation, perhaps occurring in the Iowan glacial period (See De Terra, 1949:51) in the transverse volcanic zone of central Mexico.

7. The two species are now sympatric in west central Mexico, where morphological characters (size and shape of body and length of skull) differ most. Where the two species are allopatric, these same morphological characters differ least.

8. This is a documented instance of character displacement in mammals.

9. On the basis of internal morphological characters studied (auditory ossicles, hyoid apparatus, and baculum), *Baiomys* seems to be more closely related to a South American hesperomine, perhaps *Calomys*, than to any North American cricetine.

10. Pygmy mice were more widely distributed in the past than they are at present. Part of the ancestral stock of the pygmy mice may have emigrated from North America into South America in a brief period in the Pliocene; if so, it is easy to understand why certain South American hesperomines resemble *Baiomys*.

11. The combination of morphological and behavioral characters in the living pygmy mice warrants generic status for them. If *Baiomys* were

treated as a subgenus of the genus *Peromyscus,* there would be adequate justification for including in the genus *Peromyscus* a number of other genera, some of them occurring in South America. Such lumping of genera would reduce our understanding of the natural relationships among this group of cricetine rodents.

THE END

Made in United States
Cleveland, OH
06 November 2026

47654663R00063